Power of the Shade

"Not Your Grandfather's Shrooms"

An Introductory Guide to Fungi

By

Thomas Guttry

Contents

Dedications

Dedicated to My Mom, Delynn Guttry,
for Her Strength in Softness

About the Author

The author spent his career as a television and film producer. He is an avid outdoorsman with a deep curiosity about the natural world. Some of his most joyous boyhood memories involve hiking the woods of California's Sierra Nevada and exploring the Mojave Desert. His fascination with fungi began with the strange discovery of a growth inside a rotting, fallen pine tree and a petrified Artist's Conk given to him by his grandmother.

Artist's Conk
Ganoderma applanatum

Foreword

Fungi serve as the unseen champions of the forest floor and the silent architects who weave the tapestry of life. This book honors their subtle yet powerful abilities to guide a more sustainable and healthier future through their hidden potential and influence.

It celebrates mycelium networks, which connect all living beings—both visibly and invisibly—while demonstrating how these often-overlooked organisms maintain ecological balance.

This books pays tribute to professional researchers who study fungi and to amateur mycologists whose lives are devoted to unraveling fungal secrets. This extraordinary group of organisms has been extensively explored through the passion and curiosity of scientists whose unwavering commitment to scientific inquiry continues to reveal fungi's ecological importance and potential to address human challenges.

The very existence of this book owes its realization to their continuous hard work and dedication. Their research motivates us to keep delving into fungal mysteries, learning about their fascinating properties, and celebrating their remarkable discoveries.

This work also honors individuals who possess the courage to examine what lies beneath the surface and who actively seek knowledge through hands-on exploration of everyday wonders. Within the fungal kingdom lies hidden beauty, intricate complexity, and profound marvels.

I hope this book will inspire a new generation of explorers to investigate this captivating world and to discover the immense potential found in power of the shade. Understanding fungi requires ongoing exploration, and this book serves as a foundational starting point—enabling readers to deepen their appreciation for fungi's essential roles in our lives.

May this book spark new investigations, for the secrets of the fungal world remain largely undiscovered. This dedication encourages you to pursue deeper exploration and knowledge while embracing the magical nature of fungi.

Jack O'Lantern
Omphalotus illudens

POISONOUS

Chapter 1

The Hidden World of Fungi

Imagine a hidden domain beneath us that stretches across continents and fundamentally shapes the nature of life as we understand it. The fungal kingdom encompasses a vast and diverse realm that many have largely ignored and misunderstood. People often focus on mushrooms, which represent only the visible part of an enormous underground fungal ecosystem extending deep into a complex network of mycelial filaments.

The hidden realm of fungi extends far beyond damp basements and rotting wood—it pulses with life amid the brightness of rainforests, where glowing fungi create an otherworldly nighttime illumination. Even in harsh desert environments, fungi survive by forming essential partnerships, intertwining their networks with the roots of majestic redwood trees. This realm of astonishing diversity includes delicate cup fungi and imposing giant puffballs, each telling its own story.

The extent of fungal diversity presents a fantastic spectacle. Research suggests that millions of fungal species live on Earth, many of which remain unidentified and undocumented. This concealed diversity forms an extensive reservoir of potential human benefits—from innovative medical treatments to sustainable materials—yet remains largely unexplored. Our exploration of their practical applications must begin with a moment of appreciation for the wonders of this kingdom.

Let's dispel some common myths. Fungi should not be mistaken for plants that deviated from typical characteristics. Fungi belong to their own independent kingdom, separate from plants, animals, and bacteria. They lack chlorophyll and cannot photosynthesize like plants, so they have evolved different methods of obtaining nutrition. By decomposing organic matter, fungi perform essential functions that lead to nutrient recycling and support overall ecosystem health.

Envision the forest floor as a chaotic mix of decomposing foliage and decaying timber. That apparent disorder actually represents a bustling city of fungal life. The primary body of a fungus consists of mycelia—networks of thread-like hyphae that extend through organic material and release enzymes to break down complex molecules into simpler substances. The mycelium absorbs these nutrients to support continued growth. This complex process enriches the soil and nourishes plants, sustaining forest health.

Picture a tiny world where hyphae weave a fragile yet vital pattern of life through decaying wood—an intricate tapestry that remains invisible to the human eye. Many fungi serve as nature's recyclers, playing a fundamental role in maintaining ecosystem balance.

But not all fungi are decomposers. Some function as parasites, absorbing nutrients from living hosts and often causing diseases in plants and animals. Others form mutual partnerships. Certain fungal species establish symbiotic relationships that benefit plants, animals, and even other fungi. These alliances reveal both stunning complexity and critical importance for ecological stability.

Consider the desert regions of the southwestern United States. There, fungi form symbiotic connections with plants, providing them with water and nutrients in exchange for carbohydrates produced through photosynthesis. These delicate partnerships allow both the plant and the fungus to survive in otherwise hostile environments, showcasing fungi's exceptional evolutionary resourcefulness.

A journey through a tropical rainforest reveals yet another dimension of fungal life. The moist air hangs heavy above a forest floor thick with decomposing plant matter. In this lush environment, bioluminescent fungi emit a gentle, enchanting glow, transforming the undergrowth into a magical landscape. This mystical illumination is one of the many evolutionary adaptations that fungi have developed to thrive in diverse habitats—light that attracts insects, which in turn help disperse spores.

Fungal diversity extends far beyond the familiar mushrooms. Single-celled yeasts—also fungi—play a crucial role in baking and brewing. By fermenting sugars, yeasts have enabled the creation of bread, beer, and wine for thousands of years. Some molds produce powerful antibiotics like penicillin, a miracle drug that has saved millions of lives. Though many of their deeds occur out of sight, fungi continue to shape our existence in powerful and essential ways.

The fungal kingdom functions as a vast interlinked network of organisms, sharing resources and information through extensive underground mycelial threads. Often referred to as the "wood wide web," this network connects plants, facilitating the exchange of nutrients and chemical signals. When a plant experiences stress, it can send warning signals through the mycorrhizal network to alert neighboring plants of potential danger. This fungal communication system illustrates the intricate interdependence of life on Earth.

The hidden world of fungi extends beyond biodiversity and ecological function—it holds promising solutions to many modern global challenges. Through bioremediation, fungi can help clean polluted environments. They also contribute to the development of sustainable materials, making them valuable partners in building a greener future.

Exploring this remarkable kingdom is just the beginning of our understanding of its vast potential. The following chapters will examine in detail the practical applications, capabilities, and future possibilities of fungal organisms. As we journey through this secret realm, let us pause to admire its beauty, complexity, and the silent, transformative ways fungi influence our world.

Hen of the Woods
Grifola frondosa

Chapter 2

Fungal Anatomy and Physiology

Understanding the basic components of the fungal kingdom is essential for recognizing its remarkable capabilities. Fungi do not use photosynthesis, and their multicellular structure differs from that of plants and animals. They demonstrate evolutionary success through their distinctive organizational structures and survival tactics. The mycelium represents a vast network of microscopic threads—not unlike roads and cities—and serves as the vegetative foundation of fungi, constituting the core of their existence.

Mycelium consists of hyphae, which are elongated, branching filaments similar to small interwoven rootlets. These hyphae function like fungal blood vessels and nerves, transporting nutrients and signals through extensive networks. Mycelial hyphae act as microscopic transportation systems that extend through soil and plant matter, including decaying wood and the tissues of living organisms. Their delicate walls are made of chitin, which provides both rigidity and flexibility—similar to the material found in insect exoskeletons. The strength of hyphae enables them to penetrate tough substrates and unlock nutrients inaccessible to many other life forms.

The hyphal structure displays remarkable characteristics. Many fungi have hyphae divided by compartments called septa, which resemble interconnected rooms in a vast building. The pores in these septa allow cytoplasm and organelles to move between compartments, promoting efficient transport and communication across the mycelial network. Some fungi even possess continuous multinucleated hyphae that stretch unbroken over great distances—from meters to kilometers. Managing such large, complex cells presents enormous logistical challenges, highlighting fungi's exceptional adaptability.

Mycelial expansion involves more than simple spatial growth—it requires complex environmental investigation. The apical regions of hyphae, or growing tips, sense environmental cues like nutrient gradients, moisture levels, and the presence of other organisms. With this highly developed sensing ability, hyphae navigate toward favorable conditions and avoid harmful ones. This underground structure functions as a silent yet powerful foraging network, dedicated to exploration and resource acquisition.

The scale of mycelium is astonishing. Mycelial networks from a single fungal organism can span vast areas, forming intricate ecosystems that link plants, animals, and other fungi. Picture an extensive subterranean metropolis, hidden from the human eye, silently connecting all life forms in its path. Scientists have dubbed these mycorrhizal networks the "wood wide web" because they allow plants to communicate and share resources across ecosystems.

Now, let's talk about mushrooms. The visible fungal fruiting bodies we commonly encounter are merely reproductive structures—comparable to how apples serve as the fruit of apple trees. The central portion of the fungal organism, which carries out vital functions, operates as an extensive, hidden network beneath the surface. Mushrooms are temporary structures that specialize in producing and distributing spores, the fungal equivalent of seeds.

Spores exhibit extraordinary resilience, enabling them to withstand extreme temperatures, dehydration, and even radiation. These microscopic units are dispersed by wind, water, animals, or their own propulsion mechanisms. Once a spore lands in a suitable environment, it germinates to form a new hypha, eventually developing into new mycelium. The global success of fungi is largely due to their remarkable capacity for spore dispersal.

But how do fungi obtain their sustenance? Fungi are heterotrophs—they rely on other organisms for energy, unlike plants that use photosynthesis. Their feeding strategies are diverse and fascinating. Saprophytic fungi decompose dead organic materials like leaves, wood, and animal remains. By secreting enzymes that break down complex compounds, fungi drive nutrient cycling across ecosystems. These recyclers return essential nutrients to the soil, sustaining the cycle of life.

Parasitic fungi, on the other hand, extract nutrients from living hosts. Some can destroy entire trees or infect animals and plants with diseases like athlete's foot or rust. Certain parasitic species have evolved advanced strategies to invade and establish themselves within host organisms. Some fungi even manipulate host behavior—such as directing insects to specific locations that optimize spore dispersal.

Symbiotic fungi form mutually beneficial relationships with other organisms. Through associations with plant roots, mycorrhizal fungi help plants expand their root systems and improve access to water and nutrients. In exchange, plants supply fungi with carbohydrates produced through photosynthesis. This reciprocal exchange is the foundation of many ecosystems, forming a subtle but essential partnership that ensures the survival and prosperity of both species.

The scale of this symbiosis is extraordinary. Mycorrhizal partnerships underpin forest vitality and productivity across the globe. Another remarkable example of symbiosis is the lichen, which results from a partnership between a fungal organism and an alga or cyanobacterium. The fungus provides a protective structure, while the alga or cyanobacterium contributes carbohydrates through photosynthesis. Lichens are among the first colonizers of extreme environments, thriving on bare rocks and other inhospitable surfaces. Their ability to survive in harsh conditions reflects the power of fungal-algal collaboration.

Fungi also produce a variety of bioactive compounds with important medicinal properties. Penicillin stands as a powerful testament to the medicinal potential of fungal metabolites—it revolutionized modern medicine. Today, scientists continue exploring fungi in the search for new drugs to combat cancer, infections, and other diseases.

Fungal anatomy and physiology reveal intricate features that showcase their evolutionary success. From expansive mycelial networks that connect ecosystems to specialized spores that enable wide dispersal, fungi exhibit extraordinary biological diversity. Their varied nutritional strategies allow them to thrive in countless environments.

Understanding the interplay between fungal structure and function is essential for appreciating their ecological roles, symbiotic relationships, and potential for innovation in medicine and materials. Though much of the fungal world remains hidden, it offers a vast trove of untapped potential waiting to be discovered.

Oyster Mushroom
Pleurotus ostreatus

Chapter 3

The Ecological Roles of Fungi

T he ground layer of forests often goes unnoticed, yet it is a lively hub of unseen activity where numerous living organisms thrive beyond our sight. The fungal kingdom lies at the core of this hidden world, playing an essential role in the complex relationship between life and death. Imagine a planet stripped of its quiet decomposers and dedicated recyclers—where soil fertility collapses because fungi no longer exist. Without them, unchecked decomposition would transform Earth into a wasteland of dead organic matter, a powerful testament to nature's inefficiency.

Fungi serve as the hidden champions of nutrient cycling, standing as the primary decomposers in ecosystems. These organisms are the driving force behind breaking down complex organic materials from dead plants and animals, returning vital nutrients to the environment. Without their tireless efforts, Earth would be buried under mountains of decaying biomass, and life as we know it would cease to function.

The decomposition process begins with hyphae, which form an extensive mycelium network—the vegetative structure of fungi.

Hyphae release a potent mix of enzymes and catalysts that dismantle rigid substances like cellulose, lignin, and other complex polymers found in plant cell walls and animal tissues. These enzymes break down tough structures into simple molecular components that the fungus absorbs, eventually releasing them back into the environment.

Take, for example, a fallen log. To the human eye, it appears to be a lifeless, decomposing piece of wood. But beneath the surface, a dynamic interaction unfolds—a competition and collaboration among fungi, each with a unique role. Some fungi specialize in breaking down cellulose, the primary structural element of plant cell walls. Others focus on lignin, the complex polymer that gives wood its rigidity. This decomposition is not chaotic but a coordinated microbial choreography—a division of labor in which each fungus plays a specific part in reducing the log to its basic components: carbon dioxide, water, and essential mineral nutrients.

Under a microscope, this process becomes visible: fungal hyphae weave through the decaying wood like tiny, diligent workers deconstructing a massive structure. The nutrients once trapped in the log's complex molecules are released, becoming available to nourish new plants and organisms. Fungi are the vital link in the endless cycle of life, death, and renewal.

Minerals like nitrogen and phosphorus, essential for plant development, are freed during decomposition, fueling ecosystem productivity. Carbon dioxide is released into the atmosphere, reentering the global carbon cycle. This role extends beyond nutrient recycling—fungi are also key players in regulating Earth's climate. By decomposing organic matter, they influence atmospheric carbon dioxide levels, impacting global temperatures and shaping our planet's climate systems.

Fungi's influence is not limited to forests. These organisms function in every ecosystem, including grasslands, deserts, and aquatic environments. In grasslands, fungal decomposers break down plant litter, enriching the soil and supporting vegetation growth. In aquatic ecosystems, fungi break down organic material in sediments, impacting nutrient cycling and water quality. Their activity has deep and far-reaching effects that transform ecosystems at a fundamental level.

But nutrient cycling is only one piece of the puzzle. A large and diverse group of symbiotic fungi—mycorrhizal fungi—form complex associations with plant root systems. These fungi help plants extend their root reach, improving access to soil nutrients and water. In return, plants provide the fungi with photosynthetic carbohydrates. This mutualistic exchange sustains most terrestrial ecosystems.

Imagine a forest where trees and plants form a dense, impenetrable green canopy. Beneath the surface, a secret network is hard at work. Mycorrhizal hyphae create vast underground connections between the roots of countless plants. This "Wood Wide Web" enables plants to communicate and share resources. Through the fungal network, plants support one another by exchanging nutrients, helping to sustain the entire plant community. Far from passive, this network actively influences forest dynamics by mediating plant interactions and shaping ecosystem structure and function.

The relationship between fungi and plants goes beyond nutrient exchange. Mycorrhizal fungi also protect their plant hosts from harmful pathogens by producing antibiotics and other defense compounds. These symbiotic fungi help plants tolerate drought and environmental stress, increasing their chances of survival. These partnerships are intricate and deeply integrated, forming the foundation of healthy, resilient ecosystems.

Fungi fulfill many ecological roles beyond decomposition and symbiosis. Numerous species are parasitic, infecting plants, animals, and even other fungi. Some parasitic fungi cause significant crop damage, threatening food security. Others act as pathogens that cause diseases in animals and humans. While often harmful, these parasitic fungi also help regulate population dynamics in ecosystems, preventing any single species from becoming dominant and thus contributing to ecological balance.

The interaction between decomposing fungi, mycorrhizal symbionts, and parasitic species shapes ecosystems around the world. Fungi work quietly beneath our feet, influencing nutrient cycles, supporting plant health, and playing critical roles in the structure and function of plant communities.

These microscopic organisms reveal nature's intricate design— showing how death gives rise to life and how resilience is built into ecosystems. Research into fungi continues to uncover their complex mycelial networks and powerful biochemical abilities, offering profound insight into Earth's processes and life-supporting systems.

As our understanding of fungi deepens, so does our appreciation for their vital role in maintaining planetary health. These unseen engineers of the natural world remind us that some of the most important life-sustaining systems lie just beneath the surface.

Wood Ear
Auricularia auricula-judae

Chapter 4

Mycorrhizal Networks The Underground Internet

Beneath our feet lies an enormous hidden city—a subterranean metropolis that surpasses all human structures in complexity. This city is not built from steel and concrete but from delicate threads that form an intricate network spanning continents and connecting millions of inhabitants through quiet mutual relationships. This is the world of mycorrhizal networks: nature's underground internet for fungi.

The common image of mushrooms sprouting from the ground represents only the reproductive part of a much larger fungal organism. The true reach of a fungus remains hidden beneath the soil, where a vast network of hyphae—microscopic threads—forms mutualistic relationships with plant roots. These hyphae, thinner than a human hair, create a living underground web that transforms individual plants into a complex, interconnected community.

Nature's ingenuity is evident in this intricate system, often referred to as the "Wood Wide Web." It goes far beyond simply channeling nutrients and water. It is an active system that enables plant communication, resource sharing, and the transmission of distress signals. Consider a young sapling growing in the shadow of a massive oak tree. Its limited root system may struggle to access nutrients and water. But through the mycorrhizal network, this sapling is not alone. It receives essential resources from its older, more established neighbors via fungal hyphae that act as lifelines, linking it to the greater forest. Meanwhile, the oak tree benefits by extending its reach through this network, gaining access to nutrients it couldn't otherwise reach—thanks to the sapling.

The nutrient exchange facilitated by mycorrhizal networks includes essential elements such as phosphorus and nitrogen, vital for plant growth and development. These networks also transport water, helping plants survive droughts and maintaining the overall health of plant communities. During dry periods, larger, well-established plants can share their water reserves with smaller seedlings through these underground connections, increasing the ecosystem's resilience.

But the mycorrhizal network is more than just a transportation system—it also functions as an advanced communication network, akin to a biological internet. Plants send chemical messages through the network to alert nearby plants of potential dangers such as insect infestations or diseases. The fungal network relays these signals, allowing neighboring plants to activate their defenses—creating an early warning system for the plant community.

Imagine a plant under attack by herbivores. It releases volatile organic compounds into the air—airborne chemical signals that indicate distress. Simultaneously, it sends warnings through the mycorrhizal network to its neighbors. These alerted plants then begin producing toxins or reinforcing their structural defenses in preparation for the attack. This response extends beyond individual plants—the entire ecosystem shifts in structure and function. Mycorrhizal networks promote biodiversity and ecosystem resilience by enabling cooperation across species and enhancing forest health and stability. Without this underground network, ecosystems would be more vulnerable to disease, disturbances, and biodiversity loss.

The mycorrhizal network is a powerful example of symbiosis. Fungi and plants both benefit from their relationship. Plants, through photosynthesis, produce carbohydrates—their primary energy source—which they share with the fungi. In turn, the fungi enhance the plants' access to nutrients and water, strengthening their growth and survival. This continuous mutual exchange supports productivity and creates dynamic, resilient ecosystems.

The structure of mycorrhizal networks goes beyond basic resource exchange. Different types of mycorrhizal fungi form various symbiotic relationships. Some develop ectomycorrhizae, forming a sheath around plant roots that protects against pathogens. Others create endomycorrhizae, penetrating root cells to enable direct nutrient exchange. This diversity within the fungal kingdom results in robust networks that support a wide range of plant species while adapting to changing environmental conditions.

Researchers are continually uncovering new details about these networks. They study how mycorrhizal structures contribute to ecosystem stability, carbon sequestration, and the composition of plant communities. A growing area of research focuses on how human activities—such as deforestation and industrial agriculture—disrupt these networks and how we can mitigate those effects to preserve ecosystem health.

The potential applications of this research are vast. Understanding and leveraging mycorrhizal networks could transform agriculture and forestry. By integrating these natural systems into farming practices, we can enhance crop production, improve drought resilience, and reduce dependence on synthetic fertilizers and pesticides.

Imagine a future in which cultivating mycorrhizal fungi boosts agricultural sustainability and crop health. Forests could be managed for both timber production and the preservation of underground networks, ensuring long-term ecological stability. The "Wood Wide Web" reveals nature's remarkable interconnectedness and the essential role fungi play in maintaining Earth's ecosystems.

The countless subtle processes occurring underground deeply influence environmental well-being and balance. By studying mycorrhizal networks, we begin to understand the complexity of life's web and fungi's central role in planetary health. Exploration of the subterranean world unveils a story of mutual cooperation and astonishing biological intricacy—demonstrating the quiet strength of what lies in the shade.

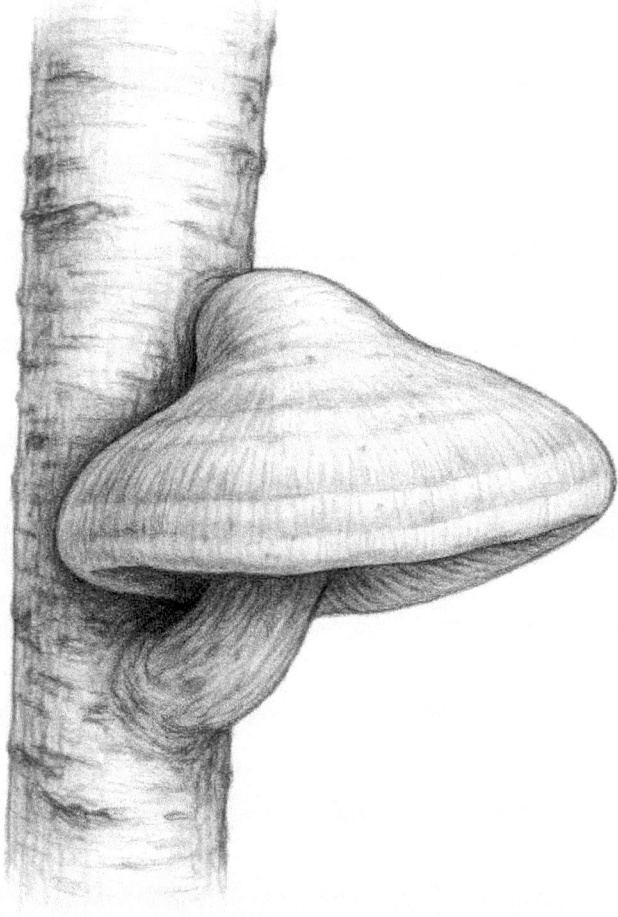

Birch Polypore
Fomitopsis betulina

Chapter 5

Fungi and Human History

A Long and Complex Relationship

Humans developed an intricate relationship with the fungal kingdom long before modern scientific understanding emerged. Our bond with fungi extended beyond mere observation to include vital elements of nourishment, healing, and spiritual connection. Fungi have deeply influenced human cultures and societies throughout the millennia in significant yet often hidden ways that continue to impact our modern world.

Food stands as one of the earliest and longest-lasting connections between humans and fungi. Research indicates that humans have harvested and consumed mushrooms for at least 6,000 years— possibly even longer. Ancient cave paintings around the world depict mushroom-like forms, suggesting early awareness of their nutritional benefits. Historical writings also highlight this connection. In the 4th century BCE, Theophrastus, the Greek philosopher and botanist, recorded various uses of fungi, studying their growth patterns and culinary value. In his *Natural History*, Pliny the Elder described

27

different mushroom species and distinguished between edible and poisonous varieties, demonstrating early knowledge of the risks involved in foraging. Mushroom gathering traditions persist in many cultures today, with local fungal knowledge passed down through generations.

Historical medical practices have further revealed fungi's significant medicinal importance. Long before modern antibiotics, traditional healing systems across cultures incorporated fungal species. In ancient Egypt, fungi were applied to wounds and used to treat various health conditions. Traditional Chinese medicine employed fungi like reishi (Ganoderma lucidum) and shiitake (Lentinula edodes), which are still recognized for their therapeutic properties today. Around the world, Indigenous communities possess deep knowledge of medicinal fungi and use them to address inflammation, infections, and other health concerns. While scientists may not fully understand the mechanisms behind these remedies, ongoing research continues to explore their active compounds and therapeutic potential. Many modern medicines have emerged from fungal sources.

Beyond food and medicine, fungi have also played crucial roles in spiritual traditions across cultures. Certain societies revered specific mushrooms as sacred, attributing mystical properties to them that connected to religious and ceremonial practices. Psychedelic fungi, such as those from the *Psilocybe* genus, were central to shamanic rituals, enabling altered states of consciousness and spiritual insight.

These secretive practices were essential for preserving spiritual beliefs and cultural identities across generations.

The sacred knowledge surrounding these mushrooms was carefully preserved within select groups through oral traditions, adding to their mystique and cultural significance. The use of such fungi involves complex rituals that require ethical and responsible oversight, acknowledging their powerful psychological effects and cultural meanings.

Fungi's influence extends far beyond their commonly recognized applications. The rise of agriculture involved complex interactions with the fungal world. Mycorrhizal fungi, for instance, play a critical role in plant growth by enhancing nutrient and water absorption. These underground partnerships have historically boosted agricultural productivity, shaping food systems throughout human development. Without these hidden fungal networks, our agricultural systems would struggle to thrive. Understanding civilization's growth requires recognition of the plant-fungi relationships that sustain our ecosystems.

The Industrial Revolution marked a turning point in human-fungal relations. Alexander Fleming's accidental discovery of penicillin ushered in the antibiotic era. Derived from the *Penicillium* fungus, penicillin revolutionized medicine by saving countless lives and transforming the treatment of infectious diseases. This breakthrough spotlighted fungi as a potent source of medical innovation and launched widespread scientific investigations into fungal species for drug development.

Following penicillin's success, researchers discovered additional fungal antibiotics, which have become indispensable in combating bacterial infections and improving global health. However, the overuse of these antibiotics has contributed to rising antimicrobial resistance—now a critical challenge in modern medicine.

Today, fungi are expanding their influence in biotechnology. They produce enzymes for industrial processes, generate sustainable biomaterials, and show promise as sources of eco-friendly biofuels. The vegetative part of fungi, known as mycelium, is gaining attention as a renewable building material that could revolutionize green construction. These diverse applications underscore fungi's versatility and potential to drive innovation across industries.

As scientific understanding of fungal biology grows, so does our appreciation of fungi's profound influence on human history. From their roles in ancient medicine and food systems to their contributions in modern technology, fungi have been vital allies—often in ways we are only beginning to understand. The bond between humans and fungi reflects both curiosity and caution, emphasizing the immense value of this often-overlooked kingdom.

The multifaceted contributions of fungi—to medicine, food security, and industrial innovation—reveal a rich narrative largely unknown to the public. Human-fungal interactions offer tremendous potential for future development, provided we approach this relationship with respect, responsibility, and sustained scientific inquiry. As we continue to explore the many roles fungi play, we uncover the intricate interweaving of life—reminding us that humans and fungi exist in an intimate, enduring relationship.

Ongoing research and ethical evaluation are essential to harness fungal capabilities without disrupting their vital ecological functions. In protecting and understanding fungi, we not only preserve a key partner in our evolutionary story—we also open new paths toward a more sustainable and interconnected future.

Button Mushroom
Agaricus bisporus

Chapter 6:

Edible Fungi:

A Culinary Delight

The scent of moist earth and decomposing leaves, usually deemed unpleasant, becomes a compelling attraction for those with refined tastes. The fragrance that leads to multiple culinary discoveries promises delicious mushrooms that can turn ordinary recipes into memorable feasts. The spectrum of textures and flavors provided by edible mushrooms, such as the delicate morel and the savory porcini, demonstrates the vast biodiversity present in the fungal kingdom.

The realm of cooking with mushrooms represents a massive and diverse culinary universe. The chanterelle mushroom, with its golden trumpet-shaped cap, shines brightly on the forest floor. Its robust structure and gentle peppery taste attract chefs who appreciate its ability to enhance straightforward dishes, revealing its natural deliciousness. While sautéed chanterelles cook, their edges become crisp in butter, creating a heady aroma that fills the kitchen with wild forest scents. A creamy chanterelle risotto emerges when the mushrooms blend perfectly with Arborio rice, producing both earthy flavors and creamy textures. Despite their basic dish appeal, chanterelles demonstrate culinary versatility by performing well in intricate sauces and gourmet preparations.

The oyster mushroom starkly contrasts the delicate chanterelle mushroom because their appearances and textures differ significantly. The larger oyster-shaped cap, along with the soft flesh of this mushroom, requires a unique cooking method. Grilling oyster mushrooms imparts a smoky char while accentuating their natural sweet taste. Their soft texture allows them to be perfect for stuffing or inclusion in stir-fries, as they easily absorb surrounding flavors. Oyster mushrooms are frequently used in vegetarian and vegan recipes because they provide a satisfying meaty texture while remaining easy to grow without sacrificing flavor or ethical standards.

Due to its meaty texture, the king oyster mushroom maintains its unique identity as a close relative. This mushroom's firm texture enables it to maintain its shape while cooking under intense heat, making it an ideal meat alternative for many recipes. You can grill marinated king oyster mushrooms to perfection and enjoy them as a delightful vegetarian kebab, or slice them and pan-fry them for an exquisite steak-alternative dish. The earthy taste of this mushroom works well with robust cheeses and thick sauces, enhancing pasta dishes, pizzas, and hearty soups.

As a renowned culinary treasure, the porcini mushroom transforms dining experiences with its unique contribution to meals.

Porcini mushrooms earned their esteemed place in Italian cooking and international menus through their deep brown caps and robust meat-like flavor. Dried porcini mushrooms deliver an intense flavor, making them essential pantry items for enhancing sauces, risottos, and soups with depth and umami. When you soak dried porcini mushrooms in hot water, their scent transports you straight to the sunlit hillsides of Tuscany. The firm texture of fresh porcini makes them ideal for grilling or sautéing, delivering a delicate flavor when they are available. Fine slicing of porcini mushrooms makes them an excellent addition to salads, providing a delicious textural contrast. The robust flavor of porcini mushrooms showcases nature's abundance with a sophisticated and rewarding taste combination.

We should now investigate less conventional edible fungi that offer exotic flavors and properties. The morel mushroom stands out for its unique honeycomb cap and subtle taste, making it an esteemed gourmet treat. This ingredient's delicate yet sophisticated nutty and earthy flavor profile makes it highly desirable for gourmet cooking. Chefs sauté morels or blend them into luxurious sauces, which demands gentle preparation methods to preserve their special texture. The short-lived springtime appearance of these fungi enhances their appeal while generating foragers' excitement and satisfaction.

The lion's mane mushroom appears with its flowing white tendrils that resemble a lion's mane. At the same time, its unexpected gentle taste adds to its distinctiveness. This cooked mushroom's texture resembles seafood, making it suitable for meals traditionally prepared with lobster or crab. Even experienced food enthusiasts will find the mushroom's surprising nature and delightful taste intriguing. Its gentle flavor makes it adaptable to various dishes, including soups, stews, fritters, and tempura. This mushroom brings culinary experiences to new, thrilling heights.

The investigation of edible fungi in culinary applications reaches beyond the study of each separate species. The gastronomic identities of global cuisines incorporate mushrooms as essential elements of their culinary traditions. Asian cooking traditions place mushrooms at the heart of dishes, which include soups, stir-fries, and complex main courses, rather than relegating them to side dishes. Japanese, Chinese, and Korean cuisines include shiitake mushrooms because their umami-rich flavor and chewy texture make them essential ingredients in noodle dishes and vegetarian dumplings. The distinct taste characteristics of these ingredients improve a wide range of recipes while showing their adaptability across multiple culinary traditions.

The historical use of mushrooms throughout European culinary practice is deeply embedded within its gastronomic traditions. The classic French mushroom duxelles serves as a flavorful base for sauces and tarts, and the highly valued Italian truffle shows how mushrooms have established their significant place in culinary traditions. The truffle demonstrates how fungi can transform basic dishes into gourmet creations through its strong aroma and earthy taste. The truffle's exceptional taste and scarcity render it synonymous with opulence and superior culinary art.

The wonderful realm of edible mushrooms demands rigorous species identification as a vital precaution. Many safe-to-eat mushrooms have poisonous doppelgangers, making mistaken consumption potentially lethal. Mushrooms should only be eaten after a knowledgeable expert confirms their identity. The dangers associated with internet research and informal mushroom identification make it crucial to avoid such methods, as the risk levels are unacceptable. This warning emphasizes the need to understand and appreciate the complex nature of fungi during exploration. Practicing responsible foraging and cooking mindfully offers more benefits than risks.

The study of edible fungi involves a combination of mycology's scientific accuracy and gastronomy's creative artistry. This exploration delves deep into taste and texture while honoring Earth's gifts and revealing how hidden places can produce exceptional flavors. The realm of mushrooms presents an extensive and intriguing culinary landscape that stretches from the forest floor to our kitchen tables. This culinary journey merges the thrill of new discoveries with the fundamental joy of enjoying a tasty meal together, proving the lasting connection between people and mushrooms. Begin your exploration of edible fungi cautiously while maintaining environmental respect, and embark on this unique culinary adventure.

The culinary treasures of mushrooms will surprise and delight every food enthusiast who explores their unique flavors and textures. The experience mirrors the diversity of mushrooms through a woven tapestry of earthy aromas and delicate flavors that traverse global culinary traditions. Culinary creativity's next groundbreaking step could lie beneath fallen leaves, waiting to be discovered and valued.

Black Trumpet
Craterellus cornucopioides

Chapter 7:

Cultivating Mushrooms at Home

E xploring the world of fungi can move beyond collecting mushrooms in nature or buying them from stores to include the satisfying practice of growing mushrooms inside your home. Turning a corner of your home or outdoor space into a flourishing mushroom farm may seem intimidating, but it is surprisingly doable for beginners. This exploration goes beyond tasting new flavors as it unveils the entire lifecycle of these amazing organisms, connecting you deeply with your food source in a meaningful way.

Selecting your mushroom species is the initial step. Oyster mushrooms are a favorite choice for beginners because their cultivation process is simple, and they grow quickly. These mushrooms easily adapt to different environments, which helps beginners cultivate them without difficulty. Shiitake mushrooms are another beginner-friendly choice with distinctive flavor characteristics. In contrast, lion's mane mushrooms offer appealing qualities to beginners with their unique appearance and delicate texture. Different mushroom species require specific substrates, with some preferring wood substrates while others grow best on straw or sawdust. Understanding these requirements is crucial for success.

After choosing your mushroom variety, the next step is acquiring the correct substrate material. Oyster mushrooms grow very well on straw as a substrate. Local farms and agricultural suppliers sell straw bales, which you can purchase. Do not use straw contaminated with pesticides or herbicides, as they can damage your mushroom culture. Proper hydration of straw before use requires soaking it in clean water for several hours or overnight to provide enough moisture for fungal growth. Excess water must be removed, as saturated substrates risk bacterial and fungal contamination. The market provides commercially pre-sterilized sawdust-based substitutes that serve as ready-to-use cultivation materials. Traditional cultivation of wood-loving species, including shiitake, relies on hardwood logs from trees like oak or maple as their substrate. Mushroom spawn is introduced into these logs to enable fungal colonization, which develops gradually.

Sterilization procedures must be performed to eliminate competing microorganisms from your substrate. Smaller substrates, such as straw, require only pasteurization. The straw must be submerged in hot water at approximately 160°F (71°C) for a designated time period to eliminate dangerous bacteria and molds. When working with large substrates, complete sterilization requires the use of pressure cooking or steam sterilization methods. Failure to properly sterilize substrates leads to a high probability of contamination and crop failure. Strict hygiene practices throughout the process are essential to prevent contamination by unwanted microorganisms.

After sterilization, the process continues with the addition of mushroom spawn. The spawn represents a cultivated culture of the chosen mushroom species that grows on grain or other appropriate media. You can purchase spawn from reputable suppliers. Sterilized substrate receives the spawn through careful mixing, which guarantees uniform distribution. When using straw as a substrate, the process requires layering straw and spawn together inside a container. Growers drill into logs to place the mushroom spawn and enable mycelium to spread throughout the wood.

The colonization phase, during which mushroom mycelium expands through its growing substrate, demands patient waiting. The time required for mushrooms to colonize the substrate can vary from several weeks to multiple months, depending on the mushroom species and environmental factors. During this phase, proper temperature and humidity levels must be maintained. Oyster mushrooms grow best when kept within a temperature range from 65 to 75 degrees Fahrenheit (18 to 24 degrees Celsius). High humidity helps promote mycelial growth. Regularly monitoring the substrate can help identify contamination indicators such as strange discolorations or rancid smells. Acting immediately to resolve problems ensures the health of your mushroom crop.

The substrate colonization process must be completed before adjusting the environment to promote mushroom fruiting. The process requires a cooler temperature and slightly reduced humidity to replicate the environmental factors that normally initiate mushroom fruiting. The colonized substrate should first be placed in a cool, dark location for oyster mushrooms before slowly adding light and fresh air. Multiple flushes of mushrooms can develop from a single substrate over a period of several weeks.

Harvesting mushrooms requires a gentle touch. Mushrooms reach harvest readiness when their caps have completely opened. Mushrooms should maintain their firmness because overripe mushrooms become mushy and lose flavor. A sharp knife is an ideal tool to harvest mushrooms because it enables clean cutting at the base. Correct harvesting methods protect the substrate from damage, allowing additional mushroom flushes. The mushrooms require gentle cleaning before they are stored properly following harvest. The ideal way to maintain the freshness of mushrooms is through refrigeration. Both oyster mushrooms and other varieties remain flavorful and retain their texture when frozen for future consumption. Mushrooms that require long-term storage benefit from drying as an appropriate preservation method. The satisfaction of building something from start to finish merges with the enjoyment of tasting the final results of your work in this experience. The whole process, from substrate preparation to mushroom harvest, teaches people to value fungi's incredible power while requiring patience and offering rewarding experiences. Your journey moves beyond your kitchen walls, leading you straight into nature's realm and the splendor of fungi. The process serves as proof of nature's strength and the fulfilling benefits of sustainable living.

Many additional mushroom species, besides oyster mushrooms, can thrive in domestic cultivation setups but will often demand specialized procedures and equipment. The cultivation of shiitake mushrooms from inoculated hardwood logs requires a more extended commitment than many other mushroom-growing methods. Lion's mane mushrooms develop their distinctive cascading form when grown on sawdust-based substrates, presenting skilled cultivators with an interesting challenge. Home mushroom cultivation remains a dynamic sector where new techniques and insights appear continuously. Engaging with seasoned mushroom growers while participating in online forums and workshops delivers essential support and guidance that will enable you to develop fungal farming skills in your personal growing space.

Home mushroom cultivation delivers culinary benefits while also presenting opportunities to achieve more sustainable practices. Home mushroom cultivation decreases the environmental burden imposed by commercial mushroom production by eliminating the need for extensive transportation and packaging. The activity promotes comprehensive knowledge about fungi's essential function in ecosystems while highlighting the vital need for ecological equilibrium and respect for nature.

Growing your own food delivers unparalleled satisfaction. Starting your homemade mushroom farm combines scientific research with attentive care and results in the great reward of harvesting your own edible mushrooms. Homegrown mushrooms offer a distinctive flavor that reflects your hard work and creates an intimate bond with nature. Mushroom cultivation provides both seasoned chefs and beginners with a rewarding journey through sustainable flavors that should occupy every kitchen. Step into the world of fungi by rolling up your sleeves and starting your culinary journey, because it brings rewards that justify the hard work.

Chaga
Inonotus obliquus

Chapter 8:

Foraging for Wild Mushrooms:

A Cautious Approach

Wild mushroom hunters are irresistibly drawn to the unexpected discovery of chanterelles hidden beneath mossy logs and the substantial feel of king boletes in their hands. Exploring the forest with a basket and knife demands both caution and respect, exceeding the requirements of a standard picnic outing. The pursuit of wild mushrooms combines human curiosity with nature's powerful—and occasionally dangerous—elements in a thrilling quest. Although the rewards, such as unique flavors and a natural connection, are substantial, they require careful consideration due to the inherent risks.

A cultivated mushroom may provide consistent results, unlike the unpredictable nature of wild mushrooms. A single misidentification of wild mushrooms can result in serious illness or even fatality. Some species possess lethal toxins, which can lead to organ failure. Wild mushrooms attract seekers due to their unpredictable nature and the unique traits each one displays as it responds to its environment. The unpredictability and uniqueness of wild mushrooms require respectful treatment combined with expert species knowledge and strict adherence to safety protocols.

The foremost and vital regulation for wild mushroom foraging is the absolute identification of the species. Anyone considering this undertaking needs to approach it with the utmost seriousness. Relying exclusively on visual matches as identification tools remains dangerously deceptive, even when field guides are employed. The appearance of many poisonous mushrooms closely matches that of safe-to-eat varieties. The death cap mushroom (*Amanita phalloides*) causes deadly confusion with edible mushroom varieties, such as specific paddy straw species. People without proper training often miss the fine distinctions between mushroom color and gill form.

An essential step for accurate mushroom identification is purchasing an extensive field guide that covers local species. These guides offer detailed descriptions and high-resolution photographs alongside microscopic characteristics, enabling a multifaceted approach to identification. The expertise and experience of a professional cannot be replicated by even the most comprehensively developed guidebook. Members of local mycological societies benefit from enhanced learning opportunities. Guided forest excursions organized by these groups provide essential learning experiences, with experienced mycologists supervising participants. Through hands-on training and myth-busting sessions, these experts offer crucial skills and knowledge that build your confidence in mushroom identification.

The use of multiple mushroom identification techniques becomes essential, aside from using field guides and expert advice. This includes careful examination of the mushroom's overall morphology: Examine the mushroom by looking at the cap's shape, size, and color; the structure of gills or pores; stem length and thickness; texture; and whether there is a ring or volva (a cup-like base structure). The mushroom's habitat information—including its associated trees or plants and soil type—helps provide essential clues. Finding a location requires thorough documentation, including taking photographs from different perspectives.

Photography serves an essential function throughout the identification process. Detailed, high-resolution photographs taken from multiple angles capture important specimen features to create an exhaustive visual document. These photographs serve as a reference when comparing with field guides or when consulting expert mycologists for verification. They function as permanent documentation, enabling you to verify your identification by retracing your steps whenever necessary.

Meticulous identification requires pairing with a conservative approach. When in doubt, throw it out. Only eat mushrooms that you have confirmed as safe for consumption. Begin with a minimal sample of the questionable mushroom to observe any allergic response. Refrain from eating more of the same mushroom until at least 24 hours have passed. Toxic mushroom consumption effects can appear many hours or even days later, so taking precautionary steps is wise.

Foraging for mushrooms requires an understanding that goes beyond collecting them, as it demands a profound respect for the natural environment. Harvest sustainably by taking only necessary amounts while leaving enough for nature and fellow foragers. Mushrooms are essential components of the forest ecosystem, serving as decomposers and symbionts and as food sources for many animals. Sustainable harvesting practices maintain the fungal population and support ecosystem wellness.

The culinary world treasures wild mushrooms because they bring distinctive and intense tastes to culinary creations, but their potent nature demands caution. Wild mushroom foraging requires slow exploration through learning and respect to achieve mindful harvesting, rather than rushing through the process. Following these safety practices while constantly learning allows you to experience the gratifying activity of wild mushroom foraging, which enriches your cooking with natural flavors and strengthens your bond with nature.

The culinary world of wild edible fungi extends far beyond the familiar chanterelles, morels, and porcini mushrooms into an expansive but often neglected realm. The saffron milk cap (*Lactarius deliciosus*) showcases delicate beauty, a vibrant orange color, and a spicy flavor that enhances pasta dishes or risotto. The black trumpet (*Craterellus cornucopioides*) adds mysterious depth to soups and stews through its dark, trumpet-shaped cap. The hedgehog mushroom (*Hydnum repandum*) displays a distinctive spiny underside and provides a nutty flavor with a hint of pepper. Meanwhile, the wood ear (*Auricularia auricula-judae*) delivers a gelatinous texture derived from elder trees, creating a contrasting effect in stir-fried dishes. The variety of mushrooms astonishes me because each species produces a distinctive culinary experience.

Traditional medicine has incorporated wild mushrooms for centuries due to their claimed benefits, which include immune system support and antibiotic effects. Conventional applications of wild mushrooms do not automatically provide scientific proof of their medicinal properties. Current research continues to explore the medical benefits of mushrooms. At the same time, experts warn against self-treatment with wild mushrooms without medical supervision. These fungi have strong chemical compounds, which make them powerful yet potentially harmful when used without proper understanding and application. Seek professional medical advice before using wild mushrooms for medicinal purposes. Wild mushroom hunting presents an opportunity for culinary exploration and serves as a way to connect deeply with the natural world through engagement with its sustaining life networks.

This experience humbles us by revealing both the strength and fragility that exist within nature's complex systems. But this power comes with responsibility. Entering this captivating realm demands a cautious approach, along with diligent identification of mushrooms and environmental respect to ensure the safety and enjoyment of the experience. Embrace the challenge, but always prioritize safety. Successful mushroom foraging delivers exceptional flavors, deep nature connections, and self-sufficiency, which justifies the hard work but requires complete respect for the dangerous mushrooms you seek. Forests contain valuable resources while hiding mysteries that should remain untouched by the inexperienced. Embark on your foraging adventures responsibly by equipping yourself with knowledge and a discerning eye while maintaining a healthy dose of caution. Respect and understanding must guide your approach toward the fungi waiting for you.

King Bolete (Porcini)
Boletus edulis

Chapter 9:

Recipes and Culinary Applications

Taking mushrooms from nature into the kitchen represents an essential phase of the mushroom journey. Once you have accurately collected your harvest, your next task is to turn these earthly treasures into delicious dishes. The following recipes demonstrate how different edible mushroom species can be used to create various dishes that bring out their unique flavors and properties. When cooking edible mushrooms, it is essential to follow proper preparation methods. Mushrooms require adequate cleaning to ensure safety, while specific cooking methods help unlock their full flavor potential.

We will begin our culinary exploration with a straightforward yet refined dish that brings out the gentle sweetness of chanterelles: Creamy Chanterelle Pasta. This cooking method suits both beginner cooks and veteran chefs perfectly. Start by slowly sautéing a pound of fresh chanterelles in butter until their aromatic oils emerge. Introduce minced garlic from one clove and a pour of dry white wine to intensify the dish's taste profile. Mix a cup of heavy cream with salt and pepper to taste, and let the sauce simmer until it reaches a slightly thicker consistency. Mix the sauce with cooked pasta, such as fettuccine or tagliatelle, and finish with grated Parmesan cheese and fresh thyme leaves. The earthy flavor of chanterelles pairs beautifully with rich cream, resulting in a perfect balance of flavor and texture. Adding a small amount of red pepper flakes imparts delightful warmth to the dish for those who enjoy a little spice.

Next, let's venture into more robust flavors with a recipe featuring the king bolete, also known as porcini: Porcini Risotto with Parmesan and Truffle Oil. Porcini mushrooms emit a potent earthy scent that captures attention. The preparation of this risotto begins with finely diced dried porcini mushrooms, which should be softened by soaking them in hot water for at least 30 minutes. Keep the soaking liquid, as it will give the dish excellent depth. Cook finely chopped shallots in olive oil in a large pan until they become tender. Toast 1 ½ cups of Arborio rice continuously for a few minutes. Slowly pour warm chicken or vegetable broth into the risotto, one ladle at a time, while stirring continuously until each batch of broth is absorbed before adding more. Achieving a creamy risotto demands patience through this careful process. When the rice approaches an al dente texture, stir in the rehydrated porcini mushrooms, soaking liquid, and generous grated Parmesan cheese. Complete the dish with a drizzle of truffle oil, followed by a sprinkle of chopped fresh parsley. The finished risotto manifests a creamy texture with earthy tastes that reflect the luxurious depth of the king bolete mushroom. Add or reduce broth until you reach your preferred risotto consistency.

Pasta dishes benefit from the saffron milk cap mushroom's vibrant orange and spicy flavor. The Saffron Milk Cap Pasta with Lemon and Herbs recipe delivers bright flavors and refreshing qualities. Begin your dish by sautéing a pound of saffron milk caps in olive oil, minced garlic, and red pepper flakes. The mushrooms' earthy, spicy flavor forms a beautiful contrast with the taste of fresh citrus. When the saffron milk caps finish cooking, add a tablespoon of lemon zest and a squeeze of fresh lemon juice, followed by a handful of chopped fresh parsley and oregano. Combine the mushroom mixture with cooked pasta, such as linguine or spaghetti, before topping it with olive oil and seasoning with salt and pepper. The vibrant, acidic taste of lemon perfectly balances the mushrooms' richness, resulting in an exceptional summer dish. The dish gains a delightful nutty flavor and texture from adding toasted pine nuts when available.

The black trumpet mushroom's versatility makes it perfect for creating a more substantial dish. The Black Trumpet Mushroom and Wild Rice Soup delivers an intense earthy taste. Begin cooking by sautéing a pound of black trumpet mushrooms with olive oil, onions, and carrots. Mix a cup of wild rice with vegetable broth into the pot. Simmer until the rice is tender. The soup gains visual appeal through the dark color of the black trumpet mushrooms. Add salt and pepper, and a hint of thyme to enhance the flavor. To complete the soup's presentation, add crème fraîche or sour cream, which provides a creamy consistency that matches the earthy taste profile. You can personalize this dish by trying vegetables like celery or parsnips and exploring various herbs to create unique variations.

The hedgehog mushroom offers a distinct culinary adventure for adventurous food enthusiasts. The Hedgehog Mushroom Tart with Gruyère Cheese brings out the mushroom's nutty and mildly peppery flavor profile. The edible spiny underside of the hedgehog mushroom gives the tart a unique textural element. Cook 8 ounces of hedgehog mushrooms with shallots and thyme until they reach a tender consistency. Mix eggs with the mushroom mixture, along with cream, Gruyère cheese, and seasonings. Fill the pre-baked tart shell with the mixture and bake until it sets. Hedgehog mushrooms deliver a nutty flavor that blends beautifully with Gruyère cheese to produce a savory tart that serves as a satisfying appetizer or light meal.

Wood ear mushrooms deliver a distinctive texture when added to stir-fry dishes. Wood Ear Mushroom Stir-fry with Ginger and Garlic creates a fast, flavorful meal. Before cooking them in a stir-fry with ginger and garlic, soak the wood ear mushrooms in warm water to soften them. Wood ear mushrooms bring a distinctive gelatinous texture to stir-fries, enhancing the dish's mouthfeel. A complete and satisfying meal can be achieved by serving it over either rice or noodles. Adding chili garlic sauce delivers more heat to this dish, which is already brimming with flavor. Wood ear mushrooms' adaptability allows them to match a broad selection of Asian dishes.

Here, we present some of the multiple culinary uses of edible mushrooms. You can find your preferred methods for cooking these interesting mushrooms through creative experimentation. Safety should be your top priority when foraging mushrooms, and you must correctly identify them before eating. Edible mushrooms span a wide array of species and forms, providing limitless culinary exploration opportunities that help deepen our understanding of nature. Embark on culinary adventures through exploration and experimentation to savor delicious outcomes. A carefully prepared and thoughtfully presented wild mushroom dish delivers an unparalleled tasting experience. The earthy tones of soil and wood, and the essence of the forest floor, create unforgettable flavors for adventurous cooks to explore. There are no boundaries to culinary possibilities except the ones you set yourself. Collect all your ingredients alongside your bravery to uncover the unique flavors residing within your kitchen.

Morel
(Morchella esculenta)

Chapter 10:

The Future of Fungi in Gastronomy

Afungal revolution stands on the brink of transforming the culinary world. Mushrooms have served as flavor enhancers in culinary history for hundreds of years, but they possess capabilities that transcend their typical function as garnishes. The mycelium network, which constitutes fungi's vegetative core, represents its essential life force and adaptability, and it will become integral to gastronomy's future. The complex system of mycelium unlocks sustainable food production, which transforms our definitions of food and alters our connection with nature.

A significant shift in food production involves the development of meat replacement products made from fungi. The fermentation process of fungi produces mycoprotein, which has quietly led the way in this industry for many years. Quorn has positioned mycoprotein as a meat substitute with high protein content and reduced fat, while maintaining a neutral taste that fits many cooking approaches. Future fungal meat substitutes will deliver even higher levels of sophistication. Scientific teams are studying different fungi and developing cultivation and processing techniques to produce textures and flavors that more accurately resemble those of animal meats. Envision a perfectly marbled "steak" cultivated from mycelium that delivers beef's satisfying chew and an earthy fungal flavor profile. Perhaps a delicate and flaky "fish" fillet grown sustainably offers a seafood alternative without the environmental drawbacks of conventional aquaculture. The potential applications within mycology are limitless, just like the fungal kingdom.

Fungi are set to transform food categories beyond meat substitutes. Fungal-based cheeses have begun to establish themselves in the dairy-free food industry. Specific fungi types can process plant milk fermentation to develop smooth-textured alternatives that match traditional cheese characteristics in texture and flavor. Fungal cheeses are an eco-friendly option that attracts vegans and dairy reducers by delivering distinct flavors beyond the typical soy or nut-based replacements. Artisanal fungal cheeses with distinctive terroirs and flavor profiles may soon enter the market as dairy cheese counterparts.

Fungi's possibilities in culinary applications allow for the creation of entirely new food products. The collective fungal biomass has become a research focus because of its potential to provide new culinary ingredients with unexplored textural and nutritional benefits. Scientific teams are working to improve fungal biomass nutrition by using genetic modification and cultivation optimization, which results in food products packed with vital vitamins, minerals, and proteins. These advanced food solutions sustainably deliver essential macro-

and micronutrients to communities facing malnutrition. Visualize high-protein fungal flour, which can be used to make bread and pastries, or nutritious fungal snacks to help fight micronutrient deficiencies in underserved areas.

Fungal gastronomy's future involves developing new ingredients and transforming established culinary practices. Precision fermentation represents a swiftly advancing field that allows controlled generation of fungal metabolites—substances created by fungi—thus presenting a substitute for conventional extraction techniques. This technology provides a controlled and sustainable approach to producing rare mushroom flavors, expanding the opportunities for gourmet cooking. Chefs could obtain truffle aroma compounds year-round from lab sources with guaranteed quality, and use precision-engineered flavors to replace those dependent on seasonal crops or slow-growing production methods.

Fungal gastronomy relies heavily on sustainable practices to foster innovative culinary techniques. Fungi show exceptional skill in converting waste streams into functional materials by transforming agricultural byproducts and organic substances. Mycelium processes complex organic matter into beneficial biomass while minimizing waste production. The process tackles waste management issues while providing a sustainable food production method that avoids harmful intensive agricultural practices. Envision urban farms of tomorrow that combine waste recycling systems with fungal cultivation to create sustainable food sources that minimize environmental harm.

Developments in technology continue to boost the expansion of fungal gastronomy. Machine learning and artificial intelligence systems enhance fungal production, producing higher yields and superior ingredient quality. Designers use three-dimensional printing technologies to fabricate intricate fungal shapes, allowing innovative presentation and design opportunities. Fungal food products offer extensive opportunities for customization and personalization. Advanced automation technologies enable the production of customized fungal dishes designed for specific dietary needs and preferences.

Despite its rapid growth, the burgeoning field faces numerous obstacles. The safety and scalability of fungal food production must be preserved for successful implementation. The safety and quality of fungal-based products depend on rigorous testing and strict quality control measures. Public acceptance and education are critical in promoting the widespread adoption of fungal foods. Addressing aversion or misconceptions about fungal foods requires an approach that combines effective communication with attractive product presentations.

Mainstream culinary adoption of fungal foods requires combined efforts from various stakeholders. To fully utilize fungi in gastronomy, we need scientists working alongside chefs, food technologists, and policymakers to solve existing challenges. The future of food depends on planetary health, and fungi present a sustainable creative direction. Adopting the multifaceted capabilities of the neglected fungal kingdom will enable us to develop a sustainable food system that remains delicious and diverse for future generations. The fungal kingdom presents earthy aromas and distinctive flavors representing enduring trends that foresee a culinary future filled with possibilities. Fungal dishes will become a key component of worldwide food systems because they bring sustainable benefits and flavorful possibilities to our tables. Gastronomy stands ready as fungi show their remarkable potency and variety to fill the palette with endless possibilities. This development represents more than a culinary revolution because it is a sustainable solution that must be tasted and appreciated.

Reishi
(Ganoderma lucidum)

Chapter 11:

Medicinal Mushrooms:

A History of Healing

Our exploration of culinary fungi is still in its infancy. However, while we discover their food potential, their historical significance as healing agents forms a much older story. Humans have identified and utilized specific mushroom species for their healing properties since before modern medicine existed. Scientific research is currently confirming traditional medicinal mushroom uses and unveiling their intricate biochemical healing processes.

Our narrative starts amidst ancient forest shadows, where mycelium fills the air with subtle vibrations to support and connect the fungal world. Particular species have become powerful partners within these concealed ecosystems, combating health disorders and weaknesses. Traditional Chinese medicine has valued Reishi (Ganoderma lucidum) for over two millennia because its shiny cap symbolizes its ability to promote longevity and enhance immune function while fighting fatigue. The earthy smell of this mushroom reveals a complex blend of triterpenes and polysaccharides, among other bioactive substances, that create therapeutic effects through their combined activity. Historical manuscripts document this substance's multiple applications, including general health tonics and targeted remedies against particular diseases.

The Japanese forests are home to another significant medicinal fungus called maitake (Grifola frondosa), which stands out because of its distinctive shelf-like formations and strong immune-boosting characteristics. The Japanese term translates to "dancing mushroom," which reflects its positive influence on health and well-being. Maitake has held a place in traditional medical practices for hundreds of years because it boosts immune function while reducing inflammation and managing blood sugar levels. The intricate branching structure equally fascinates viewers, as its biochemical composition demonstrates nature's sophisticated pharmacy system.

As we move west, we meet the Chaga fungus (Inonotus obliquus), which parasitizes birch trees and turns their bark into a dark mass resembling charcoal. Traditional medicine practitioners in Siberia and northern regions have long relied on Chaga for its strong antioxidant characteristics and claimed infection-fighting properties. Brewed from hardened conk, Chaga tea reveals its deep, earthy taste, indicating potent bioactive compounds like betulinic acid, enabling its antioxidant and anti-inflammatory properties. Nature proves its resilience through the ability of a parasitic fungus to become a powerful healing agent and source of vitality.

Medicinal mushrooms have a history that includes many more remarkable uses than these well-known examples. The mushroom known as lion's mane (Hericium erinaceus), which features white spines that look like a lion's mane, has become popular because it shows promise for enhancing brain function and encouraging nerve repair. Cordyceps sinensis is a parasitic fungus that targets insect larvae. It is essential in traditional Chinese medicine due to its energizing properties and reputed athletic performance benefits. Turkey tail (Trametes versicolor) features colorful concentric rings and is renowned for its immune system benefits and cancer support capabilities.

Traditional practices draw interest because they focus on health through a comprehensive approach. Medicinal mushrooms were used within larger wellness practices focused on healthy diets and balanced living to sustain full-body health and well-being. Traditional cultures approached fungi preparation with ritualistic methods, respecting the fungi's inherent power and nature's fragile balance. The mushrooms earned their place as sacred elements of nature, requiring reverence and meticulous attention.

Scientists today are beginning to uncover the intricate processes through which mushrooms achieve their healing benefits. Studies reveal that medicinal mushrooms contain polysaccharides with beta-glucans, which act as potent immune system modulators. These molecules enhance immune cell activity, allowing the body to combat infections and cancer cells more effectively. Research indicates that triterpenes and other bioactive compounds demonstrate antioxidant and anti-inflammatory capabilities to safeguard cells against harm and decrease inflammation. Scientific research confirms traditional knowledge and establishes a foundation for understanding fungi's healing abilities.

People must utilize medicinal mushrooms responsibly and exercise caution in their use. Before anyone uses medicinal mushrooms as part of their treatment strategy, they must secure them from trustworthy providers and seek advice from medical experts, even though these substances have been used safely for many years. Medicinal mushrooms demonstrate variable potency based on their growing conditions and processing methods, underscoring the need for stringent quality control and responsible sourcing practices. Potential interactions with prescription drugs exist, making professional oversight and informed use essential.

Extensive studies of medicinal mushrooms provide promising results and show great potential for future applications. New species and bioactive compounds discovered by scientists continuously advance our knowledge of their therapeutic uses. Through this research, we learn more than treatment options; we gain a deeper understanding of the human connection to nature and traditional knowledge, while using natural power to enhance human health.

The future of medicinal mycology is bright. Through our expanding exploration of fungi, we have discovered numerous therapeutic substances that hold the potential for treating various diseases. The path forward demands a delicate combination of traditional knowledge and respect, scientific study, and a profound comprehension of life's complex relationships. Mushrooms serve as dynamic participants within complex ecosystems, and their medicinal properties demonstrate nature's incredible complexity and adaptability. Exploring medicinal mushrooms expands our pharmaceutical knowledge and connects us to nature.

The immune cells benefit from bioactive compounds found in medicinal mushrooms that boost the body's defense against infections and cancer. Triterpenes from these compounds demonstrate anti-inflammatory and antioxidant properties that protect cells against damage while reducing inflammation. The findings reveal scientific evidence supporting the remarkable healing properties found within fungi, rather than simply repeating established traditional knowledge.

Proper caution and careful handling remain crucial when using medicinal mushrooms. Even though medicinal mushrooms have been safely used for many generations, it remains vital to obtain them from reliable vendors and consult healthcare providers about their inclusion in treatment regimens. The therapeutic effectiveness of medicinal mushrooms can vary based on cultivation and processing variables, which emphasizes the need for strict quality control and ethical sourcing standards. The possibility of medicinal mushrooms interacting with prescription medications demonstrates the critical need for professional advice.

The current investigation into medicinal mushrooms offers exciting and promising opportunities. Through continuous research, new species and bioactive substances are being revealed, which expand our understanding of their healing properties. This research seeks to develop new treatments while honoring humans' profound connection with nature through traditional knowledge and utilizing nature's abilities to improve health and wellness.

Medicinal mycology represents a field with great future possibilities. Our ongoing research into fungi reveals numerous therapeutic agents and possible treatments for various diseases. Progress in this field requires a balanced approach that honors traditional wisdom and conducts thorough scientific research while operating under an understanding of ecosystem interdependencies. The fungi we research interactively engage with complex ecosystems, and their medicinal properties reveal nature's elaborate resilience and advanced systems.

Studying their medicinal properties allows us to enhance our medical resources and strengthen our bond with nature while honoring ancient traditional wisdom. The hidden power of woodland shade reveals a dynamic system of healing resources that strengthens the long-established bond between humans and fungi, offering health benefits and sustainable solutions for future generations. The forest floor has transitioned from being seen as a simple food source to a wellspring of unexplored healing power that demands thoughtful exploration and appreciation. The exploration of medicinal mushrooms unveils a path of discovery that combines ancient knowledge with modern scientific breakthroughs and demonstrates nature's enduring healing power.

Today's scientific discoveries validate the ancient understanding of medicinal mushrooms. Contemporary laboratory research and clinical trials validate and deepen our understanding of traditional uses that have persisted through generations. The scientific community now rigorously investigates medicinal mushrooms, once only part of folklore, to understand the complex biochemical mechanisms behind their benefits.

Current scientific research emphasizes the study of polysaccharides in various medicinal mushrooms, with particular attention paid to beta-glucans. These complex carbohydrates fulfill structural roles while acting as potent immunomodulators that engage extensively with the immune system. Studies show that beta-glucans boost the performance of vital immune cells like macrophages, natural killer (NK) cells, and T lymphocytes, which play a critical role in protecting against infections and diseases. Beta-glucans operate as "signal boosters" to strengthen immune cells, helping them detect and destroy harmful agents, including viruses, bacteria, and cancerous cells. The immune enhancement represents a complex process because different beta-glucans target distinct immune pathways, resulting in multiple beneficial effects. The specific mechanisms remain under investigation, but the general advantages of increased immune alertness and strengthened protection against sicknesses stand out.

Medicinal mushrooms contain beta-glucans along with numerous other bioactive compounds. Mushrooms such as Reishi contain triterpenes, which scientists acknowledge for their ability to reduce inflammation and act as antioxidants. The body requires inflammation to respond to injuries and infections, but this response may become a harmful chronic condition if uncontrolled. Through their anti-inflammatory and antioxidant properties, triterpenes modulate the inflammatory response, reducing its intensity and duration. Triterpenes stand out among bioactive compounds due to their properties that protect cells from damage while reducing inflammation. The findings from these studies go beyond traditional beliefs and establish scientific grounds for the remarkable therapeutic qualities present in fungi.

Consumption of medicinal mushrooms should always be done with careful attention and precaution. Many medicinal mushrooms have been safely used for extended periods, but it remains vital to acquire them from reliable suppliers and involve healthcare professionals when adding them to treatment regimens. Medicinal mushroom effectiveness varies with cultivation conditions and processing methods, demonstrating the need for quality control and responsible sourcing practices. The potential for medicinal mushrooms to interact with prescription medications requires guidance from informed and professional sources.

Current research into medicinal mushrooms generates exciting possibilities for the future. Research teams continue to discover new fungal species and active substances, expanding our knowledge about their medical properties. The study targets novel therapeutic solutions and acknowledges humanity's intrinsic connection with nature, which values traditional knowledge while utilizing nature's healing potential to improve health and well-being.

Medicinal mycology represents a field with significant future prospects. Our ongoing research into fungi reveals numerous therapeutic substances and treatments for various diseases. Progress in this area requires a balanced methodology that honors indigenous wisdom alongside strict scientific research while operating within a system that considers the intricate relationships in the ecosystem. Mushrooms play an active role in their ecosystems, showcasing their natural resilience and complexity through their medicinal properties.

Through our research into medicinal properties, we build the pharmaceutical toolset and strengthen our bond with nature while honoring ancestral wisdom. Scientists have revealed how the forest's mysterious shade functions as a dynamic source of healing power, strengthening the bond between humans and fungi to ensure sustainable health benefits for future generations. Scientists and researchers have discovered that the forest floor holds significant healing potential beyond basic nutrition, which requires attentive study and appreciation. The study of medicinal mushrooms opens up paths of discovery that merge traditional knowledge with modern scientific research to demonstrate the enduring power of nature's healing properties.

Scientific developments are reviving ancient knowledge about the healing properties of medicinal mushrooms. Scientific research and clinical trials of today validate and expand our understanding of traditional medicinal mushroom uses passed down through generations. Scientific research is now thoroughly investigating these fungal allies, previously limited to folklore, by exploring the detailed biochemical mechanisms behind their positive effects.

Current scientific research primarily targets the polysaccharides present in medicinal mushrooms, with a particular interest in beta-glucans. Beyond their structural role, these complex carbohydrates also act as potent immunomodulators that interact strongly with the immune system. Studies show that beta-glucans stimulate critical immune cells such as macrophages, natural killer (NK) cells, and T lymphocytes, which are fundamental to infection defense and disease prevention. Beta-glucans work as "signal boosters," helping immune cells enhance their ability to detect and destroy viruses, bacteria, and cancer cells. The immune enhancement from beta-glucans represents a sophisticated process because different beta-glucans trigger distinct immune mechanisms, which result in various beneficial outcomes. The detailed mechanisms remain under investigation, but the benefits of heightened immune watchfulness and stronger resistance to diseases are evident.

Medicinal mushrooms contain beta-glucans along with multiple bioactive compounds with diverse properties. Mushrooms like Reishi contain triterpenes, which are well-known for their ability to reduce inflammation and fight oxidative stress. Inflammation normally protects against infections and injuries but becomes dangerous when it persists without regulation. Triterpenes help control inflammation levels and duration while also offering antioxidant support. Triterpenes stand out among these compounds because they protect cells by reducing inflammation and offering antioxidant effects. Research findings about medicinal mushrooms provide scientific validation for their unique healing qualities beyond traditional beliefs.

Medicinal mushrooms require careful usage. To ensure safety despite their historical usage, people need to obtain medicinal mushrooms from reliable suppliers and work with healthcare professionals to integrate them into treatment plans. Medicinal mushrooms demonstrate variable effectiveness depending on cultivation and processing conditions, highlighting the need for quality control and proper sourcing practices. The risk of interactions between medicinal mushrooms and prescription medications demonstrates the importance of obtaining informed professional advice.

Research into medicinal mushrooms creates an exciting and promising opportunity for future discoveries. Scientists keep discovering new species and bioactive compounds that expand our understanding of therapeutic applications. The researchers seek new medical treatments while honoring the human-nature connection through traditional knowledge use and natural healing power.

Medicinal mycology demonstrates great potential for future scientific advancements. The more profound exploration of fungi research unveils numerous therapeutic compounds that could become powerful treatments for various diseases. This development requires a balanced approach that honors indigenous wisdom through rigorous scientific research, while operating under an ecological framework that recognizes complex environmental relationships. Our studied mushrooms demonstrate active ecosystem participation, while their medicinal properties illustrate nature's complex resilience and sophisticated design.

In our ongoing research of their healing properties, we advance pharmaceutical science while deepening our bond with nature and appreciating ancestral knowledge. The forest's shade has revealed its secret power through a dynamic network of healing resources, strengthening our bond with fungi and promising future health and sustainability. What used to be perceived only as basic food now discloses an unexplored healing reservoir that demands our thorough understanding and appreciation. Exploring medicinal mushrooms leads to discoveries that blend historical knowledge with modern scientific progress, while demonstrating nature's enduring therapeutic power.

Contemporary scientific investigations are bringing fresh perspectives to ancient knowledge about the healing power of medicinal mushrooms. Modern laboratory research and clinical studies are now confirming and better explaining the traditional uses of medicinal mushrooms that have persisted through generations. Through scientific research, scientists are now exploring these medicinal fungi, which were once only part of folklore, while revealing the detailed biochemical pathways that create their positive effects.

Current scientific research primarily investigates the medicinal mushroom polysaccharides, with an emphasis on beta-glucans. These complex carbohydrates perform dual functions by providing structural support and acting as influential immunomodulators that engage with the immune system. Scientific studies show that beta-glucans can boost the functions of critical immune cells such as macrophages, natural killer (NK) cells, and T lymphocytes, which play vital roles in protecting against infections and diseases. Beta-glucans function as "signal boosters" by enhancing cellular abilities to detect and destroy viruses, bacteria, and cancer cells. The immune enhancement produced by beta-glucans represents a complex process because distinct beta-glucans interact differently with immune mechanisms to create various beneficial outcomes. The specific mechanisms remain under investigation, but researchers have identified clear benefits, including enhanced immune alertness and improved disease protection.

Medicinal mushrooms offer a wide variety of bioactive compounds beyond beta-glucans. Mushrooms like Reishi produce triterpenes, which demonstrate strong anti-inflammatory and antioxidant abilities. The body needs inflammation to heal injuries and fight infections, but it becomes dangerous when it persists beyond necessity. Triterpenes regulate inflammatory responses by reducing both their severity and duration, and they deliver antioxidant benefits. The healing potential of the shade has transitioned from its mysterious origins to become a vivid demonstration of human-fungal synergy, which will benefit future generations. The forest's natural embrace, which provided basic nourishment, now reveals its vast therapeutic potential that demands our comprehension and reverence. Investigating medicinal mushrooms is both a route of exploration and a fusion of traditional knowledge with modern scientific research, demonstrating nature's lasting strength in medicine.

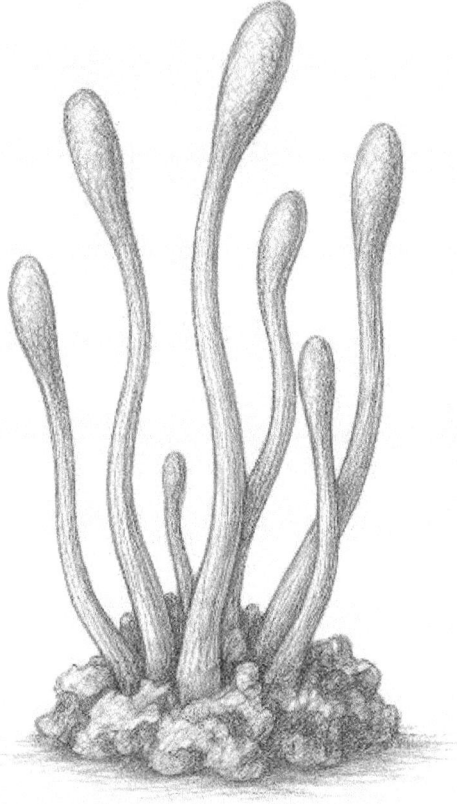

Cordyceps
(Cordyceps militaris)

Chapter 12:

Modern Applications of Medicinal Mushrooms

The powerful impact of modern science makes the ancient forests' quiet whispers hold healing secrets. Ancient practices of medicinal mushroom use are now being scientifically confirmed and better understood through modern laboratory studies and clinical research. Scientific research is exploring these once-mythic fungal allies to understand their intricate biochemical pathways that produce therapeutic benefits.

The research community continues to prioritize the study of polysaccharides in medicinal mushrooms, with beta-glucans being especially important. These complex carbohydrates function as potent immunomodulators that establish significant interactions with the body's immune system beyond their structural roles. Research findings demonstrate that beta-glucans activate the function of immune cells such as macrophages, natural killer (NK) cells, and T lymphocytes. These cells play an essential role in protecting the body from infections and illnesses. Beta-glucans work as "signal boosters" to improve their ability to identify and destroy pathogens such as viruses, bacteria, and cancer cells. The immune-strengthening action produced by beta-glucans operates through complex interactions between specific beta-glucans and various immune cells and pathways to create multiple beneficial outcomes. The precise mechanisms are still being unraveled, but the overall impact is clear: The immune system's ability to detect health threats improves while establishing a more robust defense against illnesses.

Medicinal mushrooms contain more than just beta-glucans, as they provide a complex array of bioactive substances. Mushrooms such as Reishi contain triterpenes, which have anti-inflammatory and antioxidant functions. Although inflammation serves an essential protective function during injury or infection response, it becomes harmful when it persists over time without control. Triterpenes serve as regulators of inflammation by decreasing both its strength and duration. Through their antioxidant activity, these molecules safeguard cells against damage from free radicals, unstable compounds linked to aging, and numerous diseases. Researchers are currently exploring the interaction between beta-glucans, triterpenes, and other compounds, as their combined effects likely exceed the simple sum of their individual actions. Natural systems show complex interactions, which reveal the drawbacks of studying single active compounds in isolation.

Modern medicinal mushroom research reveals exciting possibilities for their supportive role in cancer treatment. Mushrooms have not been proven to cure cancer, but research indicates they can support standard therapies by increasing treatment effectiveness and reducing side effects. Research shows that multiple mechanisms work together with cancer treatment side effects, such as inflammation and oxidative stress reduction, alongside immune modulation through beta-glucans and direct cancer cell growth inhibition. Clinical trials continue to investigate Turkey Tail mushroom as it can potentially improve chemotherapy outcomes in specific cancers through combination treatment approaches. These mushrooms should not be viewed as substitutes for standard cancer treatments, but rather as beneficial supplements that blend natural healing methods with medical precision.

Scientific investigations into medicinal mushrooms extend beyond their bodily effects because researchers examine their potential to develop new drugs and therapeutic treatments. Mushrooms feature intricate chemical structures with special molecules that can be used in medical applications. Scientists utilize sophisticated analytical techniques to both identify and isolate these compounds, while testing their effectiveness and safety for creating novel pharmaceutical products. Researchers are pursuing the fungal kingdom's biochemical diversity to discover revolutionary disease treatments, instead of reproducing existing drugs. The development of new therapies for diseases that currently have no effective treatments becomes possible through these advancements. Growing conditions, genetic differences, and specific harvesting techniques greatly influence the efficacy and quality of medicinal mushrooms.

Current cultivation techniques produce stable quality and standardized outputs, which facilitate scientific research while maintaining the consistent effectiveness of medicinal mushroom products. Establishing the credibility of medicinal mushrooms as a reliable therapeutic resource requires building trust within the industry. Switching from harvesting wild fungi to cultivating them under controlled conditions is vital in maintaining sustainability while protecting natural fungal populations from exploitation.

Mainstream medical adoption of medicinal mushrooms demands a meticulous and strategic implementation process. Scientific evidence for the therapeutic potential of these substances increases, yet further research must investigate their action mechanisms, appropriate dosages, and possible interactions with other drugs. Regulatory structures must evolve alongside scientific discoveries to maintain the safety, effectiveness, and proper labeling of medicinal mushroom products. The successful development of medicinal mushroom products requires joint efforts from researchers alongside regulatory bodies, healthcare professionals, and mushroom industry partners. Instead of replacing standard medical practices, we aim to merge traditional knowledge and modern science to create an integrated approach that enhances human health and well-being.

Exploring medicinal mushrooms reveals the sustained effectiveness of nature's pharmacy. Through this expedition, we merge ancient wisdom with modern scientific advances to reveal the abundant biodiversity surrounding us. By exploring these remarkable organisms, we uncover new treatments and gain insight into the complex relationship between humans and nature, which is essential for building a healthier and more sustainable future. The road forward demands meticulous planning, ethical acquisition of resources, and rigorous scientific research, while recognizing the exceptional ability of these organisms to deliver substantial advances in health and wellness. Medicinal mycology stands on the brink of incredible possibilities and urges us to delve into fungi's extensive yet mostly unexplored potential. This path represents both discovery and proof of ancient knowledge combined with modern science, leading to a future where forest-floor healing elements become essential to our health practices.

Cauliflower Mushroom
Sparassis crispa

Chapter 13:

Mushroom Derived Compounds and Their Benefits

S tudying the medicinal mushroom's complex chemical makeup uncovers numerous bioactive components that produce distinct therapeutic results. Beta-glucans are remarkable immune-boosting polysaccharides, which we've discussed, but their story continues beyond this point. The Kingdom of fungi contains a wealth of unique molecules that hold potential benefits for human health and wellness.

These compounds function as elite molecular athletes, focusing on specific health maintenance tasks. As established knowledge shows, beta-glucans are top immunomodulators that boost the body's natural defense mechanisms. Triterpenes comprise compounds commonly found in mushrooms, such as *Ganoderma lucidum* (Reishi). These molecules function as the body's dedicated anti-inflammatory team that suppresses the initial inflammatory reaction to injuries and infections to prevent long-term complications. The protective abilities of these compounds intensify through their antioxidant power, neutralizing harmful free radicals that trigger aging and numerous diseases. Consider free radicals as molecular vandals that damage cellular structures. At the same time, triterpenes serve as protective security guards who quickly neutralize these threats to prevent severe damage.

Sterols serve as structural components within fungal cell membranes. Although they are frequently ignored, these specific sterols, including ergosterol, show essential biological functions and sometimes work to decrease cholesterol levels. These agents perform the role of expert plumbers by managing the body's lipid profiles and maintaining crucial physiological processes. Researchers continue investigating how these compounds interact with human biochemistry while exploring their promising effects on heart health. Ongoing discoveries of novel sterols with distinct characteristics maintain a dynamic and promising landscape within this research field.

Beyond the primary compounds, medicinal fungi contain many additional bioactive substances. Short amino acid sequences known as polypeptides display diverse biological functions, such as antimicrobial and antiviral actions. These compounds act like the body's elite SWAT team, targeting and eliminating specific invaders. Certain polypeptides act as powerful antibiotics, while additional polypeptides disrupt viral replication to serve as natural antiviral compounds. Identifying and characterizing these molecules provides researchers with significant opportunities for developing new treatments against infectious diseases.

Dyeball Fungus
Pisolithus arhizus

Chapter 14:

Ethical Sourcing and Sustainable Practices

The extensive therapeutic potential of medicinal mushrooms depends on their potent bioactive compounds and ethical, sustainable acquisition practices. Collecting these fungi from sensitive ecosystems leads to profound ecological and social impacts. By neglecting these implications, we risk damaging the resources we depend on, jeopardizing future access to these precious natural medicines.

The unselective picking of wild mushrooms due to rising worldwide medicinal demand poses a significant risk to biodiversity. The maturation period of many medicinal mushrooms spans multiple years and can extend for several decades. Excessive harvesting reduces populations and interrupts the delicate ecological balance in their natural environments. Envision the forest floor as a colorful tapestry of life becoming bare after losing its fundamental fungal species. The absence of these species sends shockwaves through the ecosystem, affecting both plant life and animal populations while degrading soil quality. Particular species facing vulnerability from habitat destruction and climate shifts experience heightened extinction risks due to unsustainable harvesting practices.

The adverse effects of mushroom depletion extend beyond just the reduction of mushroom populations. Unsustainable harvesting techniques regularly result in damage to both the nearby plants and the soil structure. The reckless removal of mushrooms damages underground mycelium networks, which prevents mushrooms from growing again in the future. Mycelium networks maintain soil integrity through nutrient recycling and support forest systems by being essential components. When these networks are destroyed, they reduce future mushroom productivity while damaging the forest's health and resilience. When large quantities of fungal biomass are removed from a region, it leads to changes in nutrient cycling, which can produce ecosystem imbalances.

Harvesting wild medicinal mushrooms requires extensive travel, leading to increased carbon emissions and environmental pollution. The environmental impact increases through energy consumption in the transportation, processing, and packaging stages. The need to develop methods to reduce the ecological impact of medicinal mushroom harvesting becomes evident.

Lectins operate as sugar-binding proteins and participate in regulating immune system functions. These molecular magnets bind to specific cell surface sugars and affect cellular behavior within the immune system. These organisms demonstrate complex communication pathways with immune cells, illustrating the interconnected systems operating throughout the body. The research community's interest remains strong regarding lectins' ability to influence immune responses against cancer.

Phenolic compounds serve as potent antioxidants that protect cells by neutralizing harmful free radicals. Medicinal mushrooms contain these protective substances abundantly, which underscores their ability to defend against oxidative stress that leads to age-related diseases. These compounds boost triterpenes to enhance protection while demonstrating how nature combines multiple components to achieve healing effects.

These multifaceted compounds demonstrate their beauty through their singular properties and combined interactive effects. Medicinal mushrooms offer numerous health benefits through multiple compounds working together, rather than through the action of a single key compound. The remarkable therapeutic effect comes from the complex molecular interactions combined with the intricate interactions between beta-glucans, triterpenes, sterols, and other bioactive compounds.

The complex interaction between mushroom components highlights why researchers should study mushrooms as complete biological entities rather than focusing solely on separate compounds. Traditional methods of using whole mushroom extracts through tinctures, decoctions, and powders demonstrate an intuitive grasp of their synergistic properties. Although contemporary science enables us to separate and examine single compounds, we must uphold a holistic viewpoint, acknowledging the complex interaction between multiple elements and unidentified molecules responsible for therapeutic benefits.

Research on mushroom-derived compounds reveals their therapeutic applications are extensive and growing as scientific discoveries advance. Research demonstrates that certain mushroom extracts exhibit neuroprotective properties, promising brain health support beyond their immune modulation and anti-inflammatory effects. The antiviral potential of these natural compounds indicates their future application in fighting viral infections. These possibilities highlight the tremendous potential that natural fungal compounds hold for health and medicine, although investigations are ongoing.

Studying and learning about these compounds demonstrates why ongoing research and meticulous examination remain crucial. We require additional research to uncover how these different compounds function and establish safe dosage levels and their interactions with existing medications. Further research on these compounds must be conducted to identify any possible side effects and contraindications that will guarantee their safe and responsible use.

These compounds show promise beyond their immediate therapeutic applications. The development of novel nutraceuticals and functional foods now frequently incorporates these compounds to improve dietary products' nutritional value and health benefits. When added to foods and supplements, mushroom-derived compounds create potential opportunities for dietary health promotion and wellness improvement. This corresponds with worldwide trends prioritizing holistic well-being through nutrition and proactive healthcare strategies.

The history of medicinal mushrooms extends beyond scientific research to include traditional practices and cultural history. Medicinal mushrooms have been incorporated into the rich tapestry of human history through their use across different cultures for many generations. Medicinal mushrooms have been integral to human history across numerous cultures for centuries. Traditional knowledge is a fundamental basis for contemporary scientific research, enabling discoveries of potential therapeutic applications. Integrating traditional knowledge with modern scientific research produces an effective structure to advance new discoveries.

Indigo Milk Cap
Lactarius indigo

Medicinal mushroom research and application present vast opportunities for future scientific discoveries. Exploring fungal biochemistry leads to the discovery of unique compounds and mechanisms that offer transformative benefits for human health. Responsible development and widespread application of these powerful resources depend on ethical mushroom sourcing, sustainable cultivation methods, and strong regulatory systems.

A sustainable and effective healthcare approach emerges from merging ancient wisdom with modern science, interdisciplinary expert collaboration, and precise scientific methods. The unfolding tale of the shade-loving mushroom demonstrates the natural world's concealed capabilities, which await discovery to serve humanity. This narrative encourages us to investigate the bond between natural ecosystems and human health, while revealing how profound health solutions can be discovered in forested areas.

Society faces substantial consequences from unsustainable harvesting practices. The economic stability of numerous global communities depends on harvesting wild mushrooms. However, overharvesting drives them toward unsustainable methods, leading to additional economic challenges. The situation becomes even more worrisome for areas where residents have few other options for earning money. Wild mushroom resources face unregulated exploitation, resulting in an unfair distribution of benefits. The exclusion of local communities with expertise in sustainable harvesting and conservation efforts from decision-making processes results in social injustices.

Medicinal mushroom supply chains must transition to sustainable harvesting methods to meet ethical sourcing requirements. Sustainable mushroom harvesting requires strict collection regulations to protect populations and maintain ecological habitat balance. Successful mushroom harvesting involves expertise in the life cycles of desired species and detailed knowledge of the environments where they thrive. Practitioners who harvest responsibly utilize methods that protect the environment by picking only mature specimens while leaving enough mycelium to ensure future growth. Sustainable harvesting methods can be achieved through regulations, quotas, and community-based management systems.

Cultivation presents a potent solution for reducing dependence on natural mushroom gathering. Controlled-environment cultivation of medicinal mushrooms ensures steady production with a reduced environmental impact. Successful cultivation methods allow various species to generate high-quality medicinal mushrooms, protecting wild resources from depletion. This system guarantees a dependable medicinal fungi supply, which supports ongoing environmental sustainability and financial stability. The practice of mushroom cultivation generates economic opportunities for local communities by offering alternative sources of income and supporting natural resource conservation.

Moving toward sustainable practices demands simultaneous strategies, including adopting responsible consumption habits, implementing educational programs, and conducting scientific research. The sustainable market for medicinal mushrooms depends on consumers who actively select products sourced from sustainable methods.

Market interest in ethical products drives producers to implement sustainable production methods as consumer demand grows. Educational programs serve as crucial tools to educate people about ethical sourcing and highlight the environmental and social consequences of unsustainable harvesting methods. Such initiatives enable communities and consumers to choose wisely and engage in conservation activities. Research is fundamental in developing sustainable cultivation approaches while determining appropriate species for planting and refining production methods. Researchers are persistently investigating new methods for mushroom cultivation through the creation of sophisticated growing substrates and the refinement of environmental settings to improve yield and quality. This study enables a sustainable future where medicinal mushrooms can be grown without damaging ecosystems or affecting community welfare.

Transitioning toward sustainable practices demands cooperation between multiple stakeholders, such as governments, researchers, producers, and consumers. Combining sustainable harvesting and cultivation policies alongside consumer awareness and responsible consumption leads to a sustainable medicinal mushroom production ecosystem. Through collective efforts, we can make medicinal mushroom benefits universally available while protecting the world's forest ecosystems and their biodiversity. Our vision consists of a sustainable future where fungi's healing properties are used responsibly to maintain their availability for future generations, while protecting the environment and human needs.

The complex relationship between human health and ecological stability reveals how everything is connected. We can choose to keep the delicate balance intact or disrupt it, which will compromise the resources we depend on. Our responsibility to protect the fungal Kingdom requires us to build a sustainable future where ethical responsibilities guide our interactions with nature's pharmaceutical resources. We will only be able to fully benefit from fungi's magnificent gifts if we take responsible actions now, ensuring that future generations benefit from power of the shade.

Scarlet Elf Cup
Sarcoscypha coccinea

Chapter 15:
The Future of Mycomedicine

The forest whispers through rustling leaves and damp earth as they travel with the wind, holding secrets yet to be discovered. The medicinal possibilities fungi offer are amazing, and they are set to experience a revolutionary breakthrough that will redefine their future applications. Mycomedicine, which was previously excluded from mainstream scientific research, is now on track to become a pivotal force in transforming healthcare practices. The complex biochemical properties of mushrooms reveal potential remedies for longstanding human diseases as we enter a new chapter of scientific discovery.

Envision a time when cancer treatments no longer entail severe adverse side effects. Scientists are actively investigating how fungal compounds could improve the effectiveness of current cancer treatments. Specific mushroom-derived polysaccharides have powerful immunomodulatory effects that strengthen the body's defense against cancer cells. The goal is to improve the performance of chemotherapy and radiation by reducing their adverse side effects and increasing their therapeutic benefits. The holy grail of oncology, which involves precision targeting of cancerous cells, could be achieved by manipulating fungal-derived compounds for selective targeting. Fungal-derived compounds could destroy malignant tissue while sparing healthy cells from damage.

The therapeutic potential of mycomedicine reaches far beyond oncology, as it shows promise for treating a wide range of diseases. Fungal-based interventions represent potential solutions for neurodegenerative diseases like Alzheimer's and Parkinson's, as current treatments remain ineffective. Research demonstrates that specific fungal extracts have neuroprotective capabilities that protect brain cells from harm and may decelerate disease progression. The fungal Kingdom contains complex biochemical pathways that present numerous molecular candidates for tackling the sophisticated biological processes involved in neurodegeneration. Envision a world where cognitive deterioration from these diseases is significantly reduced, resulting in improved quality of life for countless individuals.

The persistent danger that infectious diseases pose to global health can be managed through fungal-based solutions. Antibiotic resistance escalates the urgent need to discover new antimicrobial compounds. Fungi possess many species from which researchers can derive potent antibacterial and antiviral compounds for medical use. Research into fungal byproducts demonstrates that their metabolic processes yield numerous potential drugs that scientists have yet to discover and develop. The research into fungal compounds promises to establish a new era where infections that once resisted treatment become manageable, saving lives and alleviating suffering globally.

The body's misguided immune response against its tissues defines autoimmune diseases, offering new opportunities for fungal-based treatment solutions. These fungal-derived substances demonstrate anti-inflammatory effects that help regulate the immune system's excessive activity while diminishing the symptoms and progression of debilitating diseases. Carefully manipulating these compounds could lead to therapies that target immune system dysregulation directly without weakening the body's infection-fighting capabilities.

Creating new medicinal drugs demands collaboration across multiple scientific disciplines. Successful drug development from fungal compounds requires interdisciplinary collaboration among mycologists, chemists, pharmacologists, and clinicians to discover beneficial compounds and develop safe medication treatments. The project will require multiple advanced research methods, such as genomic sequencing, metabolomics, and cutting-edge imaging technologies. High-throughput screening techniques will enable scientists to rapidly test thousands of fungal extracts, accelerating the discovery of new drugs. The development of new medicinal products rests on substantial research investments alongside strong regulatory guidelines to deliver safe and effective treatments.

The advancement of medicine goes beyond just finding new drugs. It also encompasses innovative therapeutic approaches. Therapeutic interventions using whole mushrooms or their extracts, known as mycotherapy, are earning acknowledgment for their potential therapeutic benefits. Working together, various fungal compounds could offer a broader disease management solution than drugs that contain just one molecule. A treatment model that merges ancient traditional practices with modern scientific understanding represents a potential game-changer in healthcare's transition to personalized medicine. Future medical treatments could become customized for specific diseases and individual patients by utilizing fungal extracts targeting their distinct needs.

The development and use of medicinal products require careful consideration of ethical implications. We must adopt sustainable harvesting and cultivation methods to protect wild fungal populations from depletion and preserve their natural ecosystems. The fair distribution of research benefits from mycomedicine development remains essential to ensure that those most in need receive the advantages of these discoveries. Establishing international partnerships and policy-making efforts must ensure that ethical principles become fundamental to mycomedicine's future framework.

Furthermore, public awareness and education are paramount. Public awareness about the advantages and limitations of mycomedicine helps avoid unrealistic expectations. It encourages the safe use of these powerful natural treatments. Healthcare systems must develop effective therapies while training professionals to apply biomedicine safely and effectively.

Biomedicine's future holds limitless possibilities for developing treatments to prevent and cure numerous diseases. We need a comprehensive approach that combines scientific advancement with ethical evaluation and public participation to achieve this potential. The story of mycomedicine encompasses more than scientific breakthroughs; it represents humanity's perpetual search for healing wisdom rooted in nature's complex pharmacy. The exploration continues as scientists and researchers weave together scientific precision with an ethical duty to unlock the immense possibilities within the fungal Kingdom. Exploring the fungal world provides us with new medical discoveries and reveals life's complex interconnectedness, which expands our healthcare knowledge and our perception of humanity's role in the ecosystem. The way ahead becomes clear thanks to the shining glow of bioluminescent mushrooms, which serve as a guide to a healthier, sustainable future.

Wavy Cap
Psilocybe cyanescens
PSYCHOTROPIC

Chapter 16:
Psychotropic Fungi – Wonders & Warnings

The psychotropic fungi found within the fungal kingdom attract fascination and controversy because their chemical compounds profoundly change human perception and consciousness. Throughout human history, these mushrooms have maintained a dual reputation as objects of worship and fear, being both celebrated and banned. Psychotropic fungi have sparked both religious visions and scientific curiosity while offering people intense personal revelations. These mushrooms exist within a world full of dangers and legal pitfalls while being misunderstood by many. Researching psychotropic mushrooms requires careful consideration because of their profound effects.

Consuming psychotropic fungi poses grave legal, psychological, and physiological risks that necessitate careful consideration or complete avoidance. The following chapter serves only educational purposes. This text does not instruct readers on identifying or finding psychotropic fungi or describe how to consume them. Psychotropic fungi represent mushroom species that produce psychoactive chemicals capable of modifying perception, mood, and brain functions.

Misidentifying poisonous mushrooms poses immediate physical risks, while hallucinogens can produce unpredictable psychological consequences that may become profoundly damaging. A lack of proper context and preparation during the experience creates a risk of trauma, psychosis, and dangerous behavior. Using these substances leads to legal consequences and harsh punishments in many jurisdictions because possession remains prohibited. This discussion does not serve as advice or encouragement for anyone to search for or ingest psychotropic fungi. This chapter exists purely as an educational resource, which acknowledges both the historical and biological value of its content.

Psilocybin, along with its derivative psilocin, serves as the main chemical compound that creates psychotropic effects in certain mushrooms. When they bind to serotonin receptors in the brain, they mainly target the 5-HT2A receptor, which causes alterations to sensory perception, emotional regulation, thought patterns, and self-awareness. When people take these substances, they experience intensified colors and music, while mundane objects appear mysterious or profound. During their influence, people experience temporal dilation and contraction, and the division between their self and the external world may seem to vanish completely. These experiences represent a fundamental transformation in how the brain processes information, which people find either deeply meaningful or terrifying, depending on the individual and their circumstances.

Archaeological discoveries, along with indigenous traditions, demonstrate that psilocybin mushrooms have been utilized for thousands of years. North African rock art from over six thousand years ago shows humanoid figures holding mushrooms with swirling patterns around them, indicating altered mental states. The Aztecs honored sacred mushrooms called "teonanácatl" as divine flesh during their religious ceremonies because these mushrooms enabled divine communication. In certain Mexican regions, indigenous Mazatec communities maintain ancestral practices that include psilocybin mushroom rituals directed by seasoned shamans.

Mushrooms originating from ancient times remain undomesticated and unsafe despite long-term exposure. During the twentieth century, Western scientists and adventurers brought psilocybin mushrooms back into public focus, starting with R. Gordon Wasson, who shared his experience at a Mazatec mushroom ceremony in his 1957 *Life* magazine article. During the 1960s, psychedelic experimentation became a defining feature of counterculture, leading to government actions that placed psilocybin on the Schedule I list in the United States and multiple other countries because it was categorized as a substance with high abuse potential and no medical benefits.

A significant reassessment of psilocybin has occurred in recent years. Clinical research shows that psilocybin-assisted therapy delivers significant positive outcomes for patients with depression, PTSD, terminal illness-related anxiety, and substance use disorders when administered in controlled settings. The discovery of these therapeutic benefits has reignited global interest in psychedelic treatments and driven legislative efforts toward decriminalization and medical approval in certain areas. These promising developments require careful examination and diligent attention. Psychotropic fungi serve as a potent medical tool rather than purely recreational substances.

Three psilocybin mushroom species stand out in nature: *Psilocybe cubensis*, *Psilocybe semilanceata* (the Liberty Cap), and *Psilocybe cyanescens*. *Psilocybe cubensis* holds the title of the most recognized psilocybin mushroom because of its straightforward cultivation process and potent psychoactive effects. This mushroom thrives on animal dung in warm and humid regions, especially throughout the southern United States, Central America, and Southeast Asia. *Psilocybe semilanceata*, known as the Liberty Cap, thrives in cooler climates within grassy meadows and pastures that lack heavy fertilization and cultivation. Its delicate conical cap and long slender stem make it recognizable to trained individuals but incredibly hazardous to the untrained due to toxic look-alikes.

The psilocybin and psilocin chemical levels fluctuate significantly between different mushroom species and individual specimens, as well as within separate sections of one mushroom fruiting body. The potency of substances varies according to factors like age differences, environmental conditions, and genetic variations. The unpredictable nature of wild mushrooms generates an extra layer of danger for consumers. Laboratory analysis is essential for the accurate determination of a sample's strength, as two mushrooms that appear identical might have widely different effects.

After consuming psilocybin mushrooms, users enter a "trip," which lasts four to six hours, but its aftereffects can persist well beyond this time. The onset of the experience starts with small shifts in perception, where colors gain intensity and sounds develop depth, while bodily sensations flow more smoothly. During the progression of the experience, emotions can become uncontrollably intense, memories arise with striking detail, and the perception of time completely vanishes. Reported experiences during altered states include meetings with unknown beings, the perception of complex geometric visuals, and deep sensations of connectedness to the natural world and the vast universe. The experience can become terrifying for some people as they enter cycles of anxiety or paranoia and lose their grounding through ego dissolution.

The user's mindset, along with the physical and social environment in which the experience unfolds, determines the outcome of a psilocybin journey. A stable and supportive environment promotes healing insights or spiritual awakenings, whereas chaotic and unsafe conditions can cause psychological distress. Traditional indigenous practices use experienced guides or shamans to offer structure and protection while interpreting powerful altered states to reduce associated risks. Trained therapists assume a comparable role to indigenous guides in contemporary clinical research by preparing participants thoroughly before sessions, watching over them during the process, and supporting their integration of the experience afterward.

Human error and environmental factors present the major risks in casual and uninformed psychotropic mushroom use, rather than the mushrooms themselves. The unintentional consumption of toxic mushroom species continues to represent a significant and deadly risk. The close resemblance between several deadly mushroom species and psilocybin mushrooms means that any minor misidentification could cause deadly results. People who already have mental health problems, such as those who have experienced psychosis or bipolar disorder, or who have family members with these conditions, might experience stronger negative reactions.

While research increasingly supports psilocybin's therapeutic benefits, these mushrooms continue to be illegal in numerous jurisdictions. If you possess, cultivate, or distribute these substances, you may face severe criminal charges that result in long-lasting consequences extending far beyond the personal experience itself. Legal frameworks continue to change but still show great inconsistency, maintaining harsh penalties in many regions.

The philosophical significance of the psychotropic experience exists beyond its legal and psychological implications. Changing your state of awareness should never be taken lightly. These experiences have the power to dismantle established beliefs and reveal harsh truths that may cause emotional distress. A number of people find themselves infused with new life purpose or enhanced connections to existence following the experience, while others find themselves disoriented and tormented by elusive visions that resist integration into their normal lives. In either case, the experience demands respect.

Despite their enigmatic nature and powerful effects, psychotropic fungi exist as neutral natural phenomena without moral implications of good or evil, or friend or foe. Although psychotropic fungi function as gateways to alternative perceptions of reality, they provide no assurance of attaining wisdom, healing, or enlightenment. They function as teachers in a way, but like all educators, they can show harshness or indifference while remaining mysterious. People who underestimate or treat them with arrogance could end up facing unexpected and overwhelming forces.

One must acknowledge that psychotropic fungi function according to their own intrinsic principles, which transcend human concepts of usefulness or ethical values. These organisms have adapted to survive in dim environments by consuming decaying matter and transforming dead organic material into living elements. The culinary, medicinal, and psychotropic benefits of fungi emerge from their complex, ancient existence without any intention to provide us with advantages. We explore their world as guests, rather than ruling it.

Understanding mind-altering fungi requires a foundation of knowledge combined with humility and caution. The forest paths are vast and complex, and many who explore them return altered. Only through deep respect should certain journeys begin, or be avoided completely.

Devil's Fingers
Clathrus archeri
POISONOUS

Chapter 17:

Poisonous Fungi – A Hidden Danger

The captivating strangeness and biological enigmas of fungi have long held human imagination and integrated themselves into continental folklore. Some mushrooms have fed us. Certain species of mushrooms have the power to change how we think, while also creating inspiration for various rituals. Certain fungi possess malevolent qualities that require us to both fear them and pay them respect. The fungal realm contains numerous wonders, but poisonous species stand out as a sobering testament to nature's beauty, which cannot ensure safety.

It is essential, before venturing further, to state this clearly and without reservation: A trained mycologist's direct supervision is essential during wild mushroom foraging because doing otherwise can lead to fatal consequences. The dangers are not exaggerated. Identifying mushrooms correctly is challenging because deadly fungi often look similar to safe or edible varieties. Consuming the wrong mushroom in small quantities can lead to severe sickness and possibly death or irreversible organ failure. The deadly nature of some toxins is such that vital bodily functions fail before any symptoms develop. The expertise possessed by trained professionals cannot be replicated by any book or digital resource. Anyone who doubts or wonders should adhere to the simplest of rules: When in doubt, leave it out. Recklessness leads to tragedy, while caution bears no shame.

The toxic properties of poisonous fungi emerged through natural selection to deter insects and animals rather than to harm humans. These organisms developed their chemical defenses throughout countless millennia to protect themselves against insects and animals, as well as microbial threats. Today's dangers emerge as remnants from ancient evolutionary conflicts. Fungi store their defensive toxins mainly in their fruiting bodies, which we recognize as mushrooms, while plants keep their poisonous elements hidden in roots or leaves.

Fungal toxins show diverse impacts when they interact with human physiology. Certain toxins target the liver with deadly precision. Fungal poisons work by disturbing the functioning of the kidneys and the nervous system, while others target basic cellular processes. Certain toxins trigger quick and severe gastrointestinal discomfort, while other toxins remain inactive until their effects become evident after treatment becomes ineffective. To properly understand the danger of these organisms, we need more than fear; we need to appreciate their subtle and complex nature.

Death Cap
Amanita phalloides
POISONOUS

The Death Cap mushroom *Amanita phalloides* stands out as the most notorious among poisonous fungi. The Death Cap mushroom originates from Europe but has spread worldwide, including North America and Australia, and causes most deadly mushroom poisonings. The lethal nature of these compounds, called amatoxins, comes from their ability to prevent protein production by blocking a vital enzyme that controls cellular functions. The absence of this crucial enzyme causes liver and kidney cells to die, which results in organ failure.

The Death Cap mushroom is especially dangerous because its symptoms take six to twelve hours to appear after consumption, before victims start experiencing vomiting and diarrhea. Following the deceptive recovery period, people experience false hope while irreversible damage begins to progress. Despite advanced medical treatments, the survival chance remains low because liver transplantation becomes necessary to prevent death. The already dangerous Death Cap mushroom becomes even more dangerous because it looks very similar to edible mushrooms that foragers usually encounter, like the Paddy Straw Mushroom and young Caesar's Mushrooms. Foragers with little to moderate experience might confuse a lethal mushroom with a safe edible variety. Nature's deadliest gifts do not display obvious warning signs when they make their appearance.

The equally menacing *Amanita virosa* stands as a close relative to the Death Cap and earns its grim title, "Destroying Angel." These deadly yet beautiful, ghostly white mushrooms grow throughout the woodlands and clearings of North America and Europe. They harbor the same amatoxins as the Death Cap, and their method of attack is identical: This treacherous process starts with digestive problems, then moves through a misleading calm before resulting in complete devastation.

The Deadly Galerina, also known as *Galerina marginata*, is small in size and easy to miss. These modest brown mushrooms prefer decaying wood as their growing medium, yet seem completely ordinary to people who lack specialized knowledge. Despite being harmless-looking "little brown mushrooms" to beginners, they contain deadly amatoxins and have caused fatalities, which is why experts advise new mycologists to steer clear of all small brown fungi.

Members of the *Cortinarius* genus, known as Webcaps, possess fibrous caps and rusty-colored spores that conceal a dangerous toxin called orellanine. The toxin orellanine works slowly but destructively in comparison to the rapid lethality of amatoxins. The effects of the toxin do not show up until days later when the kidneys become severely compromised. After months of dialysis, some victims recover, while many require a transplant or live their lives with compromised kidney function.

Yet not all mushroom poisons bring death. Muscarine from *Inocybe* and *Clitocybe* affects the parasympathetic nervous system, causing excessive sweat production, tear flow, excessive saliva production, nausea, and dangerously slow heart rhythms. Despite atropine administration potentially reversing muscarine poisoning when given promptly, it still poses a significant danger, especially to those who are vulnerable.

There are mushrooms that cause violent reactions, though they rarely result in death. In North America, *Chlorophyllum molybdites*, which goes by the name Green-Spored Parasol, stands as the primary mushroom that causes poisoning symptoms, including violent vomiting and diarrhea. The poisonous Jack-O'-Lantern mushroom *Omphalotus olearius* emits an eerie glow, which deceives mushroom hunters into mistaking it for chanterelles before it causes severe gastrointestinal distress. The most alarming danger from some False Morels — *Gyromitra* species — stems from their production of toxins that transform into rocket fuel chemicals inside the body, causing seizures, comas, or liver failure.

The horrible reality exists that toxic mushrooms and safe-to-eat species frequently share identical appearances. The Death Cap has edible cousins. Untrained individuals might mistake Destroying Angels for button mushrooms. Even cautious mushroom hunters seeking the flavors of spring can be deceived by False Morels. The natural world did not create these fungi to mislead humans, yet humans frequently become confused. A brief error can lock someone into an unfortunate destiny.

Mushroom poisoning manifests in many ways. Initial signs of mushroom poisoning typically consist of nausea, vomiting, diarrhea, and stomach cramps, which can start within minutes or take several hours to appear. Toxic agents trick victims into feeling recovered, only to launch a deadlier assault on their livers or kidneys afterward. These toxins rapidly damage brain and heart-lung function, which leads to hallucinations alongside seizures and breathing problems that can end in heart failure. Medical intervention needs to be obtained without delay in all instances. Delays can prove deadly. The delivery of a consumed mushroom sample enables doctors to quickly determine proper treatments.

Even though humans have accumulated knowledge for centuries, there does not exist a one-size-fits-all antidote for mushroom poisoning. Treatment is symptomatic and supportive: Doctors administer activated charcoal to restrict toxin absorption, along with intravenous fluids to treat dehydration, and medications to support heart and lung functions, while turning to organ transplants in extreme cases where organs cannot heal. Although antidotes exist for certain toxins, they remain uncommon and depend on fast, accurate diagnosis. The victim's chance of survival depends largely on the speed with which they reach out to medical professionals.

Many people find themselves tempted by the idea that mushroom foraging dangers can be eliminated by using technological tools like apps and guidebooks. This is a deadly illusion. Understanding wild fungi requires deeper analysis than just visual identification through pictures. The identification process of wild fungi involves examining microscopic structures, environmental conditions, and subtle physical characteristics that casual observers would miss. Skilled mycologists demonstrate extreme care by refusing to consume wild mushrooms unless they can identify them with absolute certainty.

Those who are truly interested in learning about the world of fungi should do so the right way: The best approach for those interested in fungi involves participating in local mycological societies and studying under experts while embracing a slow, methodical, and humble learning process. The thrill of deciphering the secrets of mushrooms demands careful awareness paired with patience. Anyone who attempts to quickly harvest wild edibles sets themselves up for disaster.

Mushrooms represent a dual nature of being gifts from nature while simultaneously presenting inherent dangers. These organisms function as healing agents while providing nourishment and generating feelings of awe. These natural elements teach us that not understanding our surroundings can lead to deadly consequences, and they show us that nature operates beyond human control. One bite based on an incorrect identification can destroy a person's entire life.

Regarding the forest as a source of amazement represents the correct and virtuous approach. Using the forest as a supermarket demonstrates a foolish approach. The knowledgeable forager understands their knowledge boundaries while recognizing that beneath the forest's beauty, there exist powerful poisons hidden under leaf litter and old logs. Exploring natural beauty and knowledge is safer than risking death from a single poisonous bite.

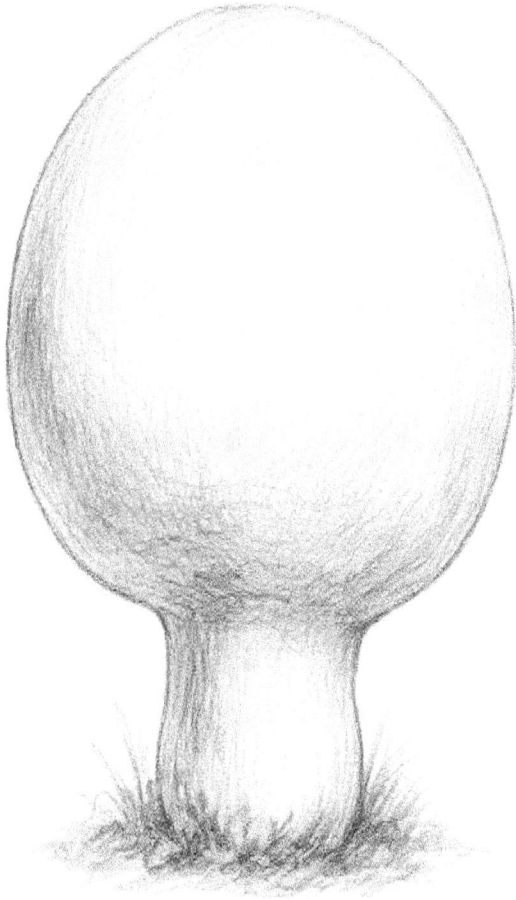

Giant Puffball
Calvatia gigantea

Chapter 18:
Fungi in Bioremediation

The forest whispers, subtle rustlings, and damp earth scents symbolize healing and restoration. Fungi's potential for medical applications fascinates people, yet its power to cleanse and restore damaged environments remains awe-inspiring. The realm of fungal bioremediation sees the humble mushroom evolve from an edible dish or medical remedy into a dedicated environmental engineer who works tirelessly to break down human waste and restore the Earth's balance.

The Earth suffers from industrial pollution, which releases toxic chemicals into the soil and poisons the planet and its living beings. Visualize an unseen army of tiny organisms whose mycelial structures spread through polluted soil to consume and detoxify contaminants into benign compounds. Fungal bioremediation demonstrates its considerable power through this process. These obscure champions operate as advanced biofactories rather than basic decomposers because they can decompose a wide variety of contaminants, including pesticides, herbicides, heavy metals, and petroleum hydrocarbons.

The complex yet beautiful processes responsible for this extraordinary capability operate through intricate systems and elegant solutions. Fungi release various enzymes and specialized biological catalysts that enable them to target particular pollutants. These enzymes act as molecular scalpels to precisely cut complex molecules into simpler, less toxic compounds. Certain fungi release extracellular enzymes into their environment, which work to break down pollutants outside their cellular structures. Fungi absorb pollutants into their cells, where intracellular enzymes work to transform these harmful substances. Regardless of the approach, hazardous waste becomes a non-toxic substance through biotransformation.

The varied nature of fungal species contributes to multiple methods of environmental cleanup through bioremediation. Each fungal species has its own specific set of enzymes that enable them to process various environmental contaminants. The enzymatic ability of white-rot fungi to degrade lignin lets them break down stubborn ecological pollutants such as polychlorinated biphenyls (PCBs) and polycyclic aromatic hydrocarbons (PAHs). Organisms with intricate metabolic systems showcase their exceptional ability to adapt and survive within toxic, chemical-dense environments that would otherwise be lethal.

Oil spills have caused severe damage to coastal ecosystems, resulting in extensive environmental destruction. Fungi show potential as effective agents for cleaning up toxic pollution sites. Some species, resilient to hydrocarbons, can convert oil components into less harmful materials, which helps speed up the natural restoration process. Fungal bioremediation provides an environmentally benign approach for cleaning up oil spills compared to conventional techniques. This method delivers improved efficiency alongside reduced costs. These remarkable organisms work silently to restore life to large oil-contaminated sand and water areas. It's not merely cleaning; it's healing.

Fungal bioremediation serves multiple environmental applications beyond just large-scale disaster response. Soil contamination from industrial activity and poor waste management represents a significant danger to human populations and natural ecosystems. Fungi restore contaminated lands to health and fertility, ensuring they become more suitable for agricultural and other land uses. The process enables land reclamation by transforming unusable areas into productive environments. The land experiences a revival of life as its toxic silence transforms into a harmonious restoration.

A persistent issue of heavy metal contamination creates opportunities for fungal bioremediation. Particular fungi demonstrate high efficiency in attaching heavy metals, which stops their environmental dispersion. Biosorption refers to the sequestration of metal ions that accumulate on fungal cell surfaces or become internalized within the fungal cells. This sequestration eliminates toxic elements from the environment so they cannot enter the biosphere. The fungal organisms function as natural filters that clean the soil and water, stopping contamination from spreading. The restoration process transcends simple detoxification through its role in restoring natural ecosystem balance.

Despite their potential for bioremediation applications, fungi face several significant challenges. Achieving optimal fungal growth and pollutant breakdown demands an in-depth understanding of environmental conditions. The effectiveness of the remediation process depends heavily on environmental factors, including temperature, pH, nutrient levels, and the specific properties of pollutants. This scientific endeavor represents an expedition toward the molecular core of ecological restoration.

A multidisciplinary approach involving mycologists, environmental scientists, engineers, and policymakers is necessary to create successful fungal bioremediation strategies. Scientific findings need practical application, which requires interdisciplinary collaboration to bridge the gap between discovery and implementation. Techniques that work well must also be cost-effective, safe for the environment, and suitable for widespread use. Moving scientific work from a controlled lab environment to real-world applications requires deliberate planning and design that addresses both ecological needs and logistical challenges. Scientific innovation works together with environmental stewardship in this partnership.

The potential of fungal bioremediation is vast. Advancing our fungal biology and ecology knowledge enables us to discover more potential in these extraordinary organisms. Advancements in genetic engineering techniques will improve fungi's bioremediation potential by enabling controlled enzyme production and increasing their resistance to extreme environmental conditions. Present opportunities for innovation indicate an upcoming era where natural healing forces lead ecological restoration efforts.

Fungal bioremediation represents a substantial opportunity to advance environmentally sustainable methods. Fungal bioremediation practices work harmoniously with circular economy principles by turning waste into valuable resources while minimizing dependence on harmful cleanup procedures. It supports ecological restoration efforts that enable damaged ecosystems to recover while boosting biodiversity. This demonstrates life's incredible adaptability through human-nature collaboration.

Environmental management strategies must include fungal bioremediation because it stands as a fundamental requirement. The expanding challenges of ecological pollution call for innovative solutions that are both sustainable and practical. We need creative, sustainable, and effective solutions. The natural pollutant-degrading ability of fungi makes them powerful allies in our mission to achieve a healthier and cleaner planet. These organisms act as hidden protectors of our planet while serving as unrecognized champions who diligently mend environmental damage caused by people. The fungal kingdom demonstrates nature's self-healing power through its story, which stands as a beacon of hope against environmental degradation. Thanks to power of the shade, we can look forward to a cleaner and brighter future.

Shaggy Mane
(*Coprinus comatus*)

Chapter 19:

Mycoremediation Techniques and Applications

Using fungi for environmental cleanup tasks displays a captivating overlap between mycology and ecological science. Mycoremediation is a dynamic field that provides sophisticated and sustainable answers to humanity's critical environmental issues. Mycoremediation functions through the extraordinary metabolic abilities of fungi, which produce numerous enzymes that can decompose complex organic and inorganic molecules. Fungi possess a specialized enzymatic toolkit that enables them to survive in toxic settings while converting harmful pollutants into safer materials. Dangerous materials are selectively broken down through a complex biochemical process that transcends basic decay.

The white-rot fungus stands out as a leading organism in mycoremediation. These organisms earn their name from their capacity to break down lignin, a complex structural element in wood. Enzymes that degrade lignin also attack resistant environmental contaminants such as toxic polychlorinated biphenyls (PCBs) and oil spill-related polycyclic aromatic hydrocarbons (PAHs). White-rot fungi demonstrate high effectiveness because they produce strong oxidases, including lignin peroxidase and manganese peroxidase, which start a series of reactions that fully mineralize these harsh pollutants. Imagine the potential: These fungi transform toxic wastelands into habitable landscapes through their silent metabolic activities. These organisms display remarkable metabolic flexibility, which enables them to survive and prosper in hostile environments that destroy other species.

The mycoremediation toolbox benefits from the contributions of multiple fungal species that extend beyond white-rot fungi. Brown-rot fungi display impressive cellulose degradation skills despite their limited recognition as decomposers of complex pollutants that affect plant matter and cause soil instability when contaminated. These organisms' significant function in repairing soil structure while enhancing nutrient recycling should receive proper recognition. Some fungal species have developed ways to endure heavy metal presence while storing these metals to serve as natural filters in polluted settings. The fungal biosorption process sequesters metal ions by attaching them to cell walls or moving them inside cells, which stops their spread and limits their availability to other living organisms. This remediation method passively yet assertively makes heavy metals inert.

Mycoremediation techniques demonstrate a wide range of applications that mirror the variety found within fungi. Special fungal strains work to break down lingering hydrocarbons in oil spill aftermath to speed up the natural restoration of damaged environments. This remediation approach has already been put into practice and produced encouraging results. Experimental tests proved that fungal-based treatments speed up oil breakdown in soil and water while minimizing the ecological damage from such disasters. Current research produces more effective fungal strains targeting oil spill cleanup operations.

Mycoremediation addresses the significant environmental problem of soil pollution due to past industrial operations and poor waste management practices. Fungal treatments help restore contaminated lands by enhancing their fertility and health, turning them into suitable areas for farming or redevelopment. More than cleaning up the mess, this process restores ecological balance and transforms lands into productive states. Mycoremediation shows promise in agricultural applications by aiding the restoration of soil that has been depleted through intensive farming methods. The strategic application of particular fungal species enhances soil nutrients, which supports plant growth and sustainable crop yield improvement. The restoration process includes soil repair while reviving the entire ecosystem's life force.

Cleaning up heavy metal contamination that endangers human health and the environment reveals promising potential through mycoremediation techniques. Specific fungi demonstrate an extraordinary capacity to capture heavy metals, which stops them from entering water systems and limits plant absorption. These mechanisms attach metal ions to fungal cell surfaces and transport them into fungal cells to extract them from their environment. The process involves more than isolating toxins; it eliminates them from the atmosphere. By optimizing environmental factors like pH and nutrients, scientists can improve fungal biofilter performance, which maximizes pollutant absorption capacity.

The use of mycoremediation presents several difficulties. Successful pollutant degradation through fungal action depends on understanding the intricate relationships between fungal biology and the environmental context of pollutants. The success of mycoremediation projects depends heavily on environmental factors, including temperature levels, pH conditions, and nutrient supply availability. Mycoremediation requires a multidisciplinary strategy that combines mycologists' knowledge with environmental scientists', engineers', and policymakers' expertise to achieve effective and sustainable results.

Advancing mycoremediation approaches requires additional research and development to achieve stronger efficiency. Critical research directions involve discovering new fungal strains with superior pollutant degradation functions while optimizing growth techniques and developing adaptable strategies for different environmental settings. The application of genetic engineering methods shows considerable potential for boosting fungal enzyme production and increasing fungal resilience to extreme environmental conditions, leading to enhanced efficiency in breaking down pollutants. A deeper understanding of fungal biology and ecology will lead to new and inventive uses of mycoremediation.

Applying mycoremediation within core environmental management practices should evolve from theoretical possibility to essential implementation. The fight against increasing pollution and environmental damage requires us to develop sustainable solutions that work effectively. The natural ability of mycoremediation to eliminate pollutants while restoring ecosystems showcases its appeal as an ecologically responsible cleanup method. This method supports circular economy standards by turning waste into usable resources and lessening our reliance on harmful technological systems. The technique creates a balanced collaboration between human inventive prowess and nature's natural regenerative processes.

Mycoremediation stands out as a strong and environmentally friendly solution for ecological repair. The exceptional metabolic abilities of fungi, combined with continuous progress in mycological research and biotechnology, present vast opportunities to resolve numerous environmental issues. This development goes beyond technological progress because it demonstrates nature's ingenious power as a subtle yet significant force for ecological restoration. Forest whispers, which symbolized healing, now provide the foundation for building a future with better environmental health. Our world undergoes transformation through the power of shade.

Lobster Mushroom
Hypomyces lactifluorum

Chapter 20:
Fungal Based Biomaterials

Beyond their application in mycoremediation, fungi demonstrate exceptional metabolic versatility. The remarkable metabolic versatility of fungi demonstrates a significant yet underexplored capacity for creating sustainable biomaterials. Mycelial network-derived fungal materials from extensive underground ecosystems present a robust substitute for conventional materials that frequently harm the environment. The combination of strength with flexibility, insulation properties, and biodegradability in these materials has the potential to transform the construction and packaging industries, along with other sectors.

Envision an environment where construction materials grow naturally rather than being extracted from mines or created through manufacturing processes. Fungal biomaterials represent real-world possibilities rather than science fiction concepts. The vegetative part of a fungus, known as mycelium, uses its network of fine white hyphae to bind different organic substrates together. Nature's design manifests through self-assembly, resulting in strong, lightweight, yet versatile materials. The cultivation of mycelium on agricultural waste materials like hemp hurds and sawdust generates a strong and sustainable composite material.

The mycelium functions as a natural binding agent by weaving through substrate components to establish a densely interlinked network. It simulates its environmental role by binding soil particles to establish a stable ecosystem. Fungal biomaterials produced through this process exhibit multiple beneficial characteristics. The material demonstrates lightweight characteristics and remarkable strength comparable to certain plastics, but it offers superior environmental benefits.

The cellular structure of these materials gives them superior insulation qualities, making them ideal for construction and packaging applications to enhance energy efficiency. Buildings made from fungal bricks demand fewer heating and cooling resources, which substantially reduces carbon emissions.

These materials naturally decompose at the end of their life cycle because they are biodegradable, eliminating the persistent waste problems in many synthetic materials. The ability of these materials to biodegrade provides a significant benefit within circular economic systems that focus on reducing waste and optimizing resource use. Compostable materials alleviate worries about toxic byproducts forming during their breakdown process. The production of these materials remains straightforward and demands low energy consumption, which distinguishes them from conventional manufacturing methods. Instead of ending up in landfills, agricultural waste can become the primary substance that promotes mycelium growth, turning waste into useful commodities. This method decreases dependence on non-renewable resources while lowering the environmental effects of waste disposal activities.

The cultivation process requires supplying mycelium with appropriate substrates and establishing optimal environmental conditions, including temperature and humidity, within a controlled space like a simple contained area to promote growth. The mycelium quickly spreads across the substrate, binds particles together, and molds itself into the desired form. Like nature's building approach, this process uses significantly less energy than manufacturing traditional materials such as concrete and plastics.

Fungal biomaterials stand out as a sustainable substitute compared to traditional materials. The production process for these materials operates independently of fossil fuels, leading to lower carbon emissions throughout their existence. Agricultural waste utilization helps decrease landfill loads while enhancing resource management efficiency. The ability of these materials to biodegrade helps lessen their environmental footprint when they reach the end of their life. Developing these processes creates an opportunity for a circular economy that transforms waste into valuable resources and reduces dependency on non-renewable natural materials.

Fungal biomaterials have diverse applications that extend well beyond the construction industry. These materials represent a viable alternative to Styrofoam and other unsustainable packaging materials. The combination of lightweight properties and insulation capabilities makes these materials perfect for packaging delicate items, while their biodegradability eliminates environmental concerns about plastic waste. Design applications benefit from these materials because their unique texture and properties enable the creation of objects that display distinctive shapes and textures. The potential is vast.

Fungal biomaterials demonstrate diverse applications beyond their current uses. Their porous structure enables them to function as effective filters, which can be applied to water purification and air filtration systems. These materials' moisture absorption capability allows their use in humidity control systems and packaging applications, which protects sensitive products from moisture damage. Scientists are actively researching how to embed bioactive compounds, like antimicrobial agents, into fungal biomaterials to boost their functionality and develop advanced composites with enhanced application-specific performance capabilities. Fungal biomaterials could be used in medical devices to provide biocompatible and biodegradable material options. The need for biocompatible and environmentally aware healthcare solutions is on the rise, and these materials have the potential to support healing functions in the future.

Fungal biomaterials research remains in its preliminary development stage. The production of fungal biomaterials faces hurdles in scaling to industrial needs while maintaining quality standards and discovering new uses. Optimizing growth conditions and processes for fungal biomaterials requires further research alongside efforts to understand their mechanical properties and enhance their functionality for specific uses. Unlocking this technology's full potential depends on exploring various fungal species alongside different substrate types. Creating standardized testing procedures and establishing solid supply networks are essential steps toward successfully commercializing this innovative technology. Present research efforts should concentrate on achieving commercial viability for these materials by investigating production methods that minimize costs and maximize scalability. The rising awareness of environmental sustainability and increasing demand for eco-friendly materials make it clear that fungal biomaterials have transformative potential across multiple industries. The characteristics that render these materials beneficial also provide essential insights to perfect the cultivation techniques needed for industrial production.

Fungal biomaterial development represents technological advancement and a fundamental change in how we interact with nature to achieve sustainability. The success of biomimicry demonstrates how we can develop cutting-edge solutions that protect the environment by studying nature's designs. This approach echoes this book's central theme: We must acknowledge fungi's vital but frequently ignored impact on Earth's well-being and future prospects. Utilizing these exceptional organisms allows us to surpass traditional material constraints and establish a sustainable and visionary future.

A fungal brick stands as a silent marker of nature's inventive power while representing more than new materials through its embodiment of innovative thinking and sustainable living. The fungal kingdom's whispers, which were once trapped underground, indicate a sustainable future that relies on their resilient strength.

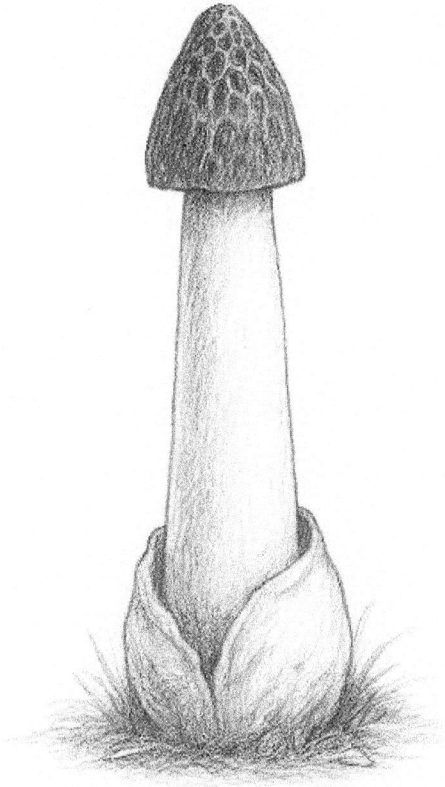

Stinkhorn
Phallus impudicus

Chapter 21:

Fungi and Waste Management

Fungi show tremendous promise as sustainable biomaterial producers, yet their influence reaches many sectors beyond construction. Their natural power to break down organic materials makes them effective partners in reducing waste. While we have examined their construction capabilities, we will now investigate their abilities to break things down. The field of fungal waste management represents a space full of novel approaches to one of the most critical environmental problems humans face.

The massive stockpiles of waste stored in landfills create a significant environmental crisis. The decomposition of organic waste within landfills happens anaerobically, leading to greenhouse gas emissions and creating visually displeasing landscapes. Methane, a potent greenhouse gas, emerges as a byproduct of this process, intensifying climate change. Landfills release dangerous chemicals that infiltrate the soil and groundwater, damaging essential natural resources. Our modern way of life produces waste at such a rate that a complete change in waste management strategies is required. Fungi present an effective solution.

Fungi's role as nature's recyclers is well established. They act as chief decomposers in multiple ecosystems, transforming complex organic molecules into simpler compounds usable by other organisms. Millennia of evolution have perfected this natural process, which we can now utilize to solve our waste management challenges. The intricate processes involved in fungal decomposition of organic substances are both complex and intriguing. Fungal-secreted enzymes function as microscale cutting tools that break down complex polymers in plant materials, plastics, and toxins by targeting cellulose and lignin. These enzymes attack specific molecular bonds in waste materials, breaking them down into basic components. This decomposition process operates through a complex biochemical cascade, demonstrating fungi's remarkable adaptive capabilities.

Fungi also show great promise in waste management through their ability to process agricultural refuse. Every year, agricultural residues—such as crop stalks, leaves, fruit peels, and other waste materials—accumulate in massive volumes, frequently ending up in landfills. These materials contribute substantially to organic waste, driving greenhouse gas emissions and environmental pollution. Fungi possess the ability to convert these waste products into valuable resources.

Through composting, fungi decompose agricultural residues into nutrient-rich compost that improves soil quality. This procedure decreases landfill waste while offering an eco-friendly substitute for artificial fertilizers. Compost resulting from fungal decomposition stands out from other composting methods due to the thorough breakdown by fungal enzymes. Fungal composting follows a straightforward process: a suitable fungal inoculum is combined with agricultural waste, and these specific fungal cultures decompose the target material under controlled temperature and moisture conditions. The fungi colonize the waste and break down organic substances into basic compounds. The resulting compost contains high levels of nutrients, improving soil fertility and promoting plant growth.

This closed-loop system converts waste materials into valuable resources in line with the principles of a circular economy. It is both efficient and effective, making it a compelling alternative to traditional composting methods, which typically require extensive time and space for decomposition. Moreover, fungi demonstrate potential applications for waste management beyond just agricultural byproducts.

Researchers have begun exploring fungi as potential agents for decomposing plastic waste—a widespread environmental contaminant. Some fungal species can break down specific types of plastics, though typically at a slower rate than cellulose. Identifying fungi with improved plastic degradation abilities remains an active area of research. This ongoing work presents a major opportunity to decrease plastic waste and support sustainable environmental goals.

Although still in the early stages, this project represents an exciting direction that could yield solutions to a significant global problem.

The potential of fungi as a treatment solution for industrial waste is also under investigation. Some fungal species can metabolize toxic substances found in industrial waste streams, reducing environmental harm. The natural metabolic processes of fungi play a crucial role in bioremediation by neutralizing hazardous pollutants and helping restore contaminated sites to healthier ecosystems. Fungi are often unaffected by these toxic substances, making them an efficient and safe option. Implementing this process on-site at the contamination source eliminates the need for long-distance transport and lowers the risk of accidental release or contact. This approach is an essential step toward reducing industrial pollution while protecting human health and environmental integrity. Such techniques offer an alternative to conventional remediation methods, which often cause additional environmental damage.

Fungi-based waste management systems offer more sustainable and efficient disposal options, avoiding traditional practices such as burning or burying waste. Incineration may reduce waste volume but produces harmful emissions. Landfills release toxic chemicals that pollute surrounding soil and waterways. In contrast, fungal-based methods provide an environmentally safe solution that reduces greenhouse gas emissions and environmental pollution. This technique supports environmental welfare while also offering economic benefits by generating useful outputs such as compost and other biological materials. These systems reduce landfill dependency and can provide substantial financial advantages for municipalities and businesses alike.

To fully harness fungi's potential in waste management, further research is necessary. Scientists continue to search for novel fungal species that can break down waste more effectively while refining current methods to boost operational efficiency and scalability. Expanding our understanding of fungal biology and biochemistry will allow for more effective environmental applications. As processes are refined, fungi could be implemented on an industrial scale, enabling resource use that supports sustainable cycles.

Fungi emerge as dynamic components of ecosystems, possessing the transformative ability to revolutionize waste management. Their capabilities for organic material degradation, natural biodegradability, and minimal environmental impact make them practical tools for reducing pollution and waste. Innovative solutions will pave the way to a sustainable future, and fungi offer a path toward a cleaner, more ecologically sound world.

Ongoing exploration of fungal capabilities continues to reveal sophisticated yet powerful methods for converting waste into usable resources while establishing a circular economic model that protects the planet's fragile balance. The quiet yet robust power of the fungal revolution holds the promise of a brighter, sustainable future—where waste is no longer an obstacle but an opportunity.

Fungi serve as pivotal forces in ecological systems and demonstrate transformative potential for waste management. Their ability to decompose a wide range of organic materials, combined with their natural biodegradability and reduced environmental footprint, makes them essential instruments in pollution control and resource recovery. Our path to sustainability demands innovation, and fungi stand ready as key players in that journey. By exploring their vast potential, we can discover elegant and practical solutions that convert waste into value while maintaining the delicate balance of our planet. The powerful yet silent fungal revolution opens the door to a sustainable future, turning waste into opportunity.

Wine Cap
Stropharia rugosoannulata

Chapter 22:

The Future of Fungi in Environmental Conservation

Fungi show exceptional waste management abilities and provide essential solutions to numerous environmental challenges. Beyond waste management, fungi offer answers to major ecological problems that humanity faces, including tackling climate change and supporting biodiversity. The future of environmental conservation depends on our ability to better understand and utilize these remarkable organisms.

Fungi offer exceptionally valuable help in battling climate change. Scientists have already proven fungi's critical role in carbon sequestration. Mycorrhizal fungi function as complex root networks that actively remove carbon dioxide from the atmosphere and deposit it into the soil. This removal of greenhouse gases plays a key role in mitigating climate change, as it prevents these gases from driving global warming. Extensive mycelial networks beneath our feet serve as massive carbon storage systems that quietly yet effectively manage Earth's carbon cycle. By studying how these fungi lock away carbon and identifying ways to boost their performance, researchers can enhance carbon capture efforts. Envision a future where scientists manage fungal networks to absorb atmospheric CO_2—acting as enormous natural carbon sponges.

Specific fungal species display exceptional capabilities to decompose tough organic molecules, such as biomass components like lignin. This ability to degrade tough organic compounds is crucial for creating sustainable biofuel production methods. The process of making traditional biofuels requires considerable energy input and produces unwanted byproducts. However, biomass pre-treatment with fungi creates a more digestible substrate for biofuel-producing microbes, enhancing efficiency while minimizing environmental impact. Utilizing this methodology promises to reduce our dependence on fossil fuels and support the transition to a sustainable energy framework. Research in this field remains in its early stages but holds significant potential for future development.

Fungi demonstrate multiple environmental functions beyond carbon sequestration and biofuel production. Their natural capacity to break down various organic materials gives them potential applications in cleaning polluted environments. Fungi can decompose numerous contaminants, such as heavy metals and pesticides, leading to reduced levels of these substances in soil and water. This bioremediation approach presents a cost-effective and eco-friendly alternative to traditional methods, which are known for high energy use and poor effectiveness against challenging pollutants. Carefully cultivated fungal colonies work to restore landscapes damaged by industrial waste, helping transform these areas into thriving ecosystems. What was once considered pure science fiction is now becoming an achievable reality.

Fungi also show enormous potential as a tool for biodiversity conservation within future environmental strategies. In many ecosystems, fungi are keystone species that maintain ecological balance through their essential roles. Plants depend on mycorrhizal networks—formed through symbiotic relationships with fungi—for health and growth support. This symbiosis is critical in changing climate conditions, as fungi help plants endure environmental stress while preserving the ecosystem's fragile equilibrium. The diversity of fungal species is essential, especially since many remain undiscovered, and their roles are not yet fully understood. The elimination of fungal variety poses severe threats to the stability and resilience of ecosystems. Preserving fungal diversity is therefore a critical component of comprehensive conservation initiatives.

Protecting fungal diversity requires us to rethink and transform current environmental management strategies. Our ecological focus must expand to include the often-neglected fungal world, which holds equal—if not greater—value than macro-organisms like plants and animals. The long-term survival of fungi depends on protecting their natural habitats, such as forests and wetlands. Research into fungal ecology and evolution will help scientists better understand fungi's roles in different ecosystems. A complete understanding will allow for more effective conservation strategies. Combining scientific study with cutting-edge genetic techniques can also lead to the development of fungal strains engineered for specific environmental cleanup tasks and enhanced carbon absorption.

Fungi offer conservation benefits that go beyond carbon sequestration, bioremediation, and biodiversity protection. Increasingly, industries and manufacturers are adopting fungi-based solutions as viable substitutes for conventional materials. Mycelium-based materials provide sustainable, biodegradable replacements for plastics and other non-renewable resources. These materials are lightweight, durable, and possess excellent insulation properties, making them useful for applications such as packaging and construction. Mycelium product development continues to make impressive strides toward establishing sustainable materials as replacements for environmentally harmful substances. With scalable production growing rapidly, mycelium is becoming a commercially viable alternative.

Multiple factors will shape the future role of fungi in environmental conservation. Continued study of fungal biology, ecology, and biochemistry is essential to harness their full potential. The successful translation of research into real-world applications depends heavily on partnerships among scientists, policymakers, and industry leaders. The development of educational programs and public awareness campaigns is also vital for promoting a widespread understanding of fungi and their ecological significance. To ensure future generations inherit a healthier planet, we must globally commit to sustainable environmental practices that place fungi at the forefront. Recognizing fungi as a foundational part of nature should lead us to use them in addressing humanity's most pressing environmental challenges.

Envision a time when engineered fungal systems detoxify polluted lands and restore damaged terrains into flourishing ecosystems. Mycelium-based materials could help construct urban infrastructure while reducing dependence on concrete and other high-carbon alternatives. Converting agricultural waste into valuable resources through fungal composting would reduce landfill loads and improve soil fertility. The incredible potential of the fungal kingdom can turn these visionary goals into practical realities—if we choose to embrace them.

The future of environmental conservation depends on fungi. Their role is not just significant—it is essential. We must fully utilize fungi's vast potential to develop sustainable solutions for the well-being of our planet and future generations. To reach that future, we must unite, innovate, and gain deeper knowledge of the fungal domain, which holds countless undiscovered solutions. The future, indeed, is fungal.

Enoki

Flammulina velutipes

Chapter 23:

Fungal Enzymes And Their Applications

Beyond mushrooms populating forest floors and topping pizzas lies the extensive universe of fungi. A treasure trove of biological marvels is hidden within their microscopic structures. Fungi produce protein catalysts known as enzymes, which act as molecular workhorses—enabling essential chemical reactions for fungal survival and offering numerous industrial applications. The distinctive capabilities of fungal enzymes to operate effectively at reduced temperatures and pH levels, while breaking down complex polymers, make them highly appealing for industrial use.

Fungal enzymes demonstrate exceptional performance in food processing applications. Consider the smooth texture of your favorite bread or the sharp tang of aged cheese—fungal enzymes are essential components in the production of these culinary delights. Amylases play a crucial role in baking by converting complex starches into simple sugars, which yeast then ferments to create the fluffy, airy textures we enjoy. Enzymes from diverse fungi are carefully regulated to ensure superior fermentation processes and to maintain product quality standards.

During cheese ripening, fungal enzymes such as lipases and proteases collaborate to develop flavor and texture. Fatty acids released by lipase activity contribute to complex aromas, while proteases break down proteins to alter texture and taste. The specific fungal enzymes selected for cheesemaking often represent proprietary knowledge that defines the unique characteristics of different cheese varieties. Fungal enzymes also aid in winemaking by breaking down pectin, helping to create smoother alcoholic beverages.

Fungal enzymes are influencing textile manufacturing far beyond traditional culinary applications. The denim industry, once reliant on harsh chemical treatments, is gradually shifting toward fungal enzyme-based processes as environmentally friendly alternatives. Various fungal species produce cellulases, which are used in bio-stoning to soften and fade denim while reducing the need for damaging chemicals. This shift results in decreased environmental impact.

These enzymes are also applied in textile resizing (removing sizing agents), scouring (cleaning fabrics), and bleaching processes. Their biodegradability, along with reduced energy demands, makes fungal enzymes attractive options for improving sustainability in the textile supply chain. The industry is increasingly adopting these bio-processing methods as ongoing research uncovers new enzymes that further enhance efficiency and eco-friendliness.

Fungal enzymes also play a pivotal role in biofuel production due to their multifunctionality. As previously discussed, fungi are essential agents in decomposing complex plant structures to unlock their energy potential. Fungal enzymes significantly improve the biomass pretreatment phase, which has traditionally been a major bottleneck in biofuel production. Enzymes such as cellulases, hemicellulases, and ligninases work together to break down plant cell walls, releasing carbohydrates that are then fermented into biofuels.

Unlike chemical alternatives, fungal enzymes function under milder conditions, resulting in lower energy consumption and fewer harmful byproducts. Scientists are actively researching ways to discover and develop fungal enzymes with enhanced durability and performance. The ability of these enzymes to increase biomass conversion efficiency represents a major breakthrough, potentially making renewable fuels economically viable for mainstream use.

Beyond biofuels, fungal enzymes have wide-ranging industrial applications. In the paper industry, they improve pulp quality while reducing reliance on harmful chemicals. In pharmaceuticals, fungal enzymes help synthesize various drugs. They also aid environmental remediation by breaking down harmful substances such as pesticides and heavy metals.

Several key fungal enzymes deserve closer examination. The cellulase enzyme group degrades cellulose—Earth's most abundant organic polymer and a structural component in plant cell walls. Efficient cellulose breakdown is essential for many industrial processes, including textile production and biofuel generation. This process involves three main enzymes: endoglucanases, exoglucanases, and β-glucosidases. Endoglucanases cleave internal bonds within cellulose chains to produce shorter fragments. Exoglucanases then remove cellobiose units (pairs of glucose molecules) from the ends of these chains, and β-glucosidases convert cellobiose into individual glucose molecules for use in various industrial processes. Together, these enzymes achieve complete and effective cellulose degradation.

Pectinases are another essential group of enzymes that degrade pectin—a complex polysaccharide in plant cell walls—used in many food processing applications. Pectinases are critical in juice clarification, winemaking, and the production of fruit-based products. By breaking down pectin, these enzymes increase both the yield and quality of fruit juices and wines. Pectinolytic enzymes include polygalacturonases, pectin lyases, and pectin methyl esterases, each targeting different pectin structures. Selecting the right combination of these enzymes is key to achieving optimal results in specific food processes.

Ligninases are fungal enzymes that break down lignin, a tough polymer that binds cellulose and hemicellulose in plant cell walls. While lignin's complex nature makes it difficult to degrade, certain fungi can produce ligninases that do so effectively. These enzymes are instrumental in biofuel production and the pulp and paper industries by granting access to cellulose and hemicellulose for conversion into valuable products. Scientists are also exploring ligninases for environmental cleanup, as they can degrade a wide range of pollutants.

With recent advancements in genetic and protein engineering, scientists can now create engineered enzymes with improved activity, stability, and specificity. These innovations broaden the scope of fungal enzymes across multiple sectors. Researchers continue to develop more efficient and sustainable enzyme technologies, driving innovation and helping industries move toward greener alternatives.

As the global search for eco-friendly solutions intensifies, fungal enzymes will play an increasingly vital role. Their remarkable versatility and environmental advantages position them as powerful tools for building a more sustainable and responsible future across food, textile, fuel, pharmaceutical, and environmental industries.

Fungi will define the future as we uncover their vast, untapped potential. Fungal enzymes demonstrate complex functionality that opens new pathways in biotechnology, contributing to a more sustainable future. Medicinal mushrooms contain bioactive compounds that boost immune cell activity and support the body's defense against infections and cancer. Among these, triterpenes stand out for their anti-inflammatory and antioxidant properties, protecting cells and reducing inflammation. Current findings provide scientific evidence of fungi's remarkable healing abilities—advancing beyond conventional knowledge.

Medicinal mushrooms must be used carefully and responsibly. Despite their long history of safe use, it is important to source them from reliable suppliers and consult healthcare professionals when incorporating them into treatment plans. Their therapeutic effectiveness varies based on cultivation environments and processing techniques, making stringent quality control and responsible sourcing essential. Additionally, the potential for interactions between medicinal mushrooms and prescription medications requires informed, professional guidance.

Current investigations into medicinal mushrooms offer a promising and exciting future. Scientific exploration continues to reveal new species and bioactive compounds, expanding our understanding of their medicinal potential. This research not only aims to discover new treatments but also celebrates humanity's deep connection with nature—respecting traditional wisdom while leveraging nature's gifts to improve health outcomes.

Medicinal mycology represents a field with enormous potential for advancement. Continued exploration of fungi reveals a wealth of therapeutic substances that may treat a wide range of diseases. Progress in this field depends on an integrated approach that blends scientific research with respect for traditional knowledge, all while considering delicate ecosystem interactions. Mushrooms function dynamically within their environments and exhibit natural healing abilities, reflecting the intricate resilience and sophistication of the natural world.

Through ongoing research into their medicinal properties, we enhance pharmaceutical resources while deepening our connection to nature and honoring ancient knowledge. The forest's shade reveals a hidden network of healing potential—demonstrating how humans have long been intricately connected to fungi in ways that may ensure future health and sustainability. Once seen merely as food, the forest floor now unveils a powerful healing force that we are only beginning to understand and appreciate.

The study of medicinal mushrooms continues to deliver surprising discoveries as ancient knowledge integrates with modern science, validating nature's medicinal power. Contemporary scientific research brings new life to the traditional wisdom surrounding medicinal mushrooms. Remedies passed down through generations now gain expanded understanding and empirical support through laboratory studies and clinical trials. What was once considered folklore is now being investigated for the complex biochemical interactions responsible for these mushrooms' health benefits.

Modern research primarily focuses on polysaccharides found in medicinal mushrooms—especially beta-glucans. These complex carbohydrates serve as powerful immunomodulators, going beyond their structural roles to engage deeply with the immune system. Scientific studies show that beta-glucans stimulate critical immune cells such as macrophages, natural killer (NK) cells, and T lymphocytes, all of which play essential roles in protecting the body from infections and diseases.

Beta-glucans act as "signal boosters," enhancing immune cells' ability to detect and destroy threats like viruses, bacteria, and cancer cells. This immune enhancement is a multifaceted process, activating several mechanisms to produce a range of beneficial effects. While researchers continue to investigate the exact pathways involved, it is clear that beta-glucans increase immune surveillance and bolster the body's defense systems.

In addition to beta-glucans, medicinal mushrooms contain numerous bioactive compounds that support health. For example, Reishi mushrooms are rich in triterpenes, known for their potent anti-inflammatory and antioxidant effects. While inflammation is necessary for healing injuries and fighting infections, chronic inflammation can be harmful. Triterpenes help regulate inflammation by reducing its intensity and duration while also protecting cells from oxidative stress. Their dual action makes them particularly valuable for maintaining cellular health and reducing damage caused by prolonged inflammation.

Modern research continues to validate traditional uses by offering scientific explanations for fungi's medicinal benefits. Still, people should use medicinal mushrooms with caution. Although many have a long track record of safe use, it is critical to ensure proper sourcing from reputable providers and to seek guidance from healthcare professionals when integrating them into treatment plans. Their effectiveness can vary depending on cultivation and processing methods, underscoring the need for rigorous quality control and responsible use. The potential for interactions with prescription medications further emphasizes the importance of professional supervision.

The exploration of medicinal mushrooms presents a compelling opportunity filled with potential benefits. Ongoing scientific investigation is uncovering new species and compounds that broaden our understanding of fungal healing properties. This research not only aims to discover novel treatments but also honors the deep, time-honored relationship between humans and nature. By acknowledging traditional knowledge and utilizing nature's healing power, we enhance human health and well-being.

Medicinal mycology remains an exciting and rapidly growing area of research. Continued exploration of fungi has already revealed numerous therapeutic compounds and potential treatments for various illnesses. Advancing this field requires a harmonious integration of traditional insights and scientific discovery, grounded in an appreciation for ecosystem dynamics. As mushrooms continue to demonstrate their ecological importance and healing potential, they offer a powerful example of nature's complexity, resilience, and enduring wisdom.

Deadly Galerina
Galerina marginata

POISONOUS

Chapter 24:
Pharmaceuticals Derived from Fungi

The extraordinary versatility of fungi extends beyond their enzymatic applications to include pharmaceutical uses. Throughout history, humans have unknowingly utilized fungi for their medicinal properties to address diverse health problems. Modern biotechnological advancements showcase the vast potential of these organisms and open new pathways for developing innovative, fungi-derived pharmaceuticals. This emerging field offers abundant opportunities as scientists harness the intricate biochemical properties of fungi to tackle critical human health challenges.

The development of antibiotics from fungi is a prime illustration of fungal-based pharmaceuticals. Penicillin is a definitive example—an antibiotic that became renowned for controlling bacterial infections. In 1928, Alexander Fleming accidentally discovered penicillin from the Penicillium fungus genus, revolutionizing modern medicine and earning him a Nobel Prize. Through its ability to stop harmful bacteria, penicillin and its many derivatives have saved countless lives by enabling the effective treatment of severe infections, including bacterial pneumonia and syphilis.

The discovery of penicillin launched a focused search for additional fungal-derived antibiotics, leading to the identification of life-saving drugs such as cephalosporins. These antibiotics attack multiple components of bacterial cell walls and metabolic pathways, providing powerful treatment options. This exploration of fungal resources produced numerous antibiotics that transformed infectious disease treatment. Thanks to evolving knowledge and technologies, the extraction and purification of antibiotics now utilize advanced industrial methods that ensure optimal yield and purity—far beyond the basic techniques of earlier years.

Fungi also represent new frontiers for life-saving medications beyond their established role in antibiotic production. Researchers extract cyclosporine, a potent immunosuppressant drug, from the fungus Tolypocladium inflatum. This drug has revolutionized organ transplantation by significantly decreasing rejection rates. Cyclosporine enhances transplant recipients' survival by suppressing immune responses to foreign tissues. It operates through complex mechanisms, targeting calcineurin—an essential enzyme in immune response pathways—highlighting the biochemical sophistication of fungi.

Statins represent another major pharmaceutical class developed from fungi. Lovastatin, the first marketed statin, was extracted from the oyster mushroom fungus Pleurotus ostreatus. Statins have dramatically improved cardiovascular health by lowering the risk of heart attacks and strokes. These drugs work by blocking HMG-CoA reductase, a key enzyme in cholesterol production. Statins show that even simple organisms like fungi can produce powerful medical compounds with substantial global health benefits.

Creating fungi-based pharmaceutical products represents a compelling intersection of scientific research and technological progress. Initially, drug production relied on cultivating fungi in large fermentation tanks to obtain active compounds. These early efforts yielded minimal quantities, and extraction and purification were labor-intensive and costly. Modern biotechnology has transformed this process: genetic engineering allows scientists to enhance fungal strains to boost specific compound production. These improvements have made drug manufacturing more efficient and cost-effective. Advances in downstream processing techniques have also improved purification, resulting in higher-quality pharmaceuticals.

Producing drugs from fungi offers multiple advantages. Fungi are easy and affordable to cultivate. Under controlled conditions, they grow in large volumes with less energy and resource consumption than other methods. Their natural ability to generate therapeutic compounds reduces the need for complex chemical synthesis. Fungi-based production not only cuts costs but also supports environmentally friendly manufacturing. Additionally, fungal biomass degrades naturally, minimizing the environmental footprint of pharmaceutical production.

Fungi produce vital drugs that save lives beyond their well-known antibiotic roles. As mentioned, the fungus Tolypocladium inflatum is the source of cyclosporine, the immunosuppressant that has revolutionized organ transplantation by greatly reducing organ rejection risks. By suppressing immune reactions to foreign tissue, cyclosporine provides millions of transplant recipients with a second chance at life. Its complex mechanisms target and inhibit calcineurin, vital to immune function, showcasing the intricacy of fungal biochemistry.

Statins, too, have become widely used for lowering cholesterol. Lovastatin, the first to be marketed, came from Pleurotus ostreatus (the oyster mushroom). These medications have drastically improved cardiovascular outcomes by reducing heart attack and stroke risks. Their mechanism of action involves blocking HMG-CoA reductase, thereby inhibiting cholesterol synthesis. Statins illustrate how fungi, despite their simplicity, can produce powerful treatments that have profoundly improved human health around the world.

Fungal-based pharmaceutical production exemplifies the synergy between science and technology. The initial stages involved growing fungi in vast fermentation tanks, from which compounds were extracted and purified—an expensive and inefficient process. Today, thanks to genetic engineering, fungal strains can be tailored to increase yields of desired compounds. This has made the process significantly more efficient and cost-effective. Innovations in downstream processing have improved purification techniques, maintaining high pharmaceutical standards.

This production method presents several advantages. Fungi cultivation is both simple and economical. They grow rapidly in controlled environments, requiring less energy than many conventional techniques. Because fungi naturally produce the needed compounds, complex synthetic chemistry is often unnecessary. Fungal drug production supports both cost savings and eco-friendly practices. Moreover, fungal biomass biodegrades easily, reducing the environmental impact of the process.

However, challenges remain. Researchers are still working to optimize fermentation conditions and improve downstream processing methods. Exploring fungal biodiversity to discover new therapeutic compounds remains an active and promising field. The complexity of fungal biochemistry offers both immense opportunity and significant scientific hurdles. Ongoing research is needed to fully understand the mechanisms of action, efficacy, and potential side effects of these compounds. As sustainability in pharmaceutical production becomes increasingly important, ensuring the environmental responsibility of fungal cultivation is a key concern.

The future of fungal-based pharmaceuticals is bright. Researchers continue to study new fungal species and develop modern techniques to identify and synthesize novel medicinal compounds. Advances in genomics and metabolomics are helping scientists decode the pathways fungi use to produce bioactive substances. This paves the way for more efficient and affordable production. The growing threat of antibiotic resistance has further accelerated interest in fungal therapies for their promising ability to treat infections. Fungal-derived medicines represent an emerging field with vast, untapped potential. Today's research in fungal biochemistry and biotechnology lays the groundwork for fungi to play a central role in future medical advancements.

Medical breakthroughs from fungal research will enhance human health in ways we are only beginning to imagine. The fungal kingdom holds vast, untapped potential that will help address future medical challenges. As medicine evolves, advanced technologies combined with continued exploration of fungal diversity will establish fungi-based pharmaceuticals as essential tools in the global healthcare arsenal.

Chicken of the Woods
Laetiporus sulphureus

Chapter 25:

Fungal Genomics and Genetic Engineering

Fungal biotechnology advances rely on finding natural compounds and genetic engineering methods that transform fungi into highly efficient biofactories. The analysis of fungal genomes through fungal genomics forms the essential basis for this modern engineering achievement. The sequencing and analysis of fungal DNA have enabled scientists to access extensive information about fungal genetic structures. Through this understanding, scientists can precisely alter fungal genetic structures to improve current characteristics and add entirely new features.

A fungal strain may generate a valuable enzyme in nature yet produce it insufficiently. Improving production through traditional methods requires careful selection and breeding processes, which take several years and provide minimal control over results. Genetic engineering offers a superior method that delivers better precision and efficiency. Scientists use CRISPR-Cas9 technology to perform precise edits on fungal DNA to achieve accurate modifications. They can now target genes controlling enzyme production to boost their activity, significantly raising production yields. The outcomes achieved through modern techniques outperform traditional methods in speed and efficiency while offering superior control.

CRISPR-Cas9 functions as precise molecular scissors by cutting DNA at designated spots. The DNA cut activates the cell's repair systems, which scientists can control by inserting desired modifications, such as adding a gene for improved enzyme function or boosting existing gene activity. CRISPR-Cas9's precision enables us to overcome traditional methods, which tend to create random and unpredictable genetic modifications. Researchers can significantly reduce unintended effects and achieve dependable, predictable outcomes by focusing specifically on target areas. CRISPR technology's effectiveness and straightforward nature have created new opportunities for genetic modification in fungi, accelerating advancement across multiple applications.

Fungal biotechnology also employs genetic engineering methods other than CRISPR-Cas9. The Agrobacterium-mediated transformation technique is a proven approach for the genetic insertion of foreign genes into fungal cells. The bacterium Agrobacterium tumefaciens functions as a natural genetic engineer by transferring its DNA into plant cells. The genetic modification of Agrobacterium enables it to transport desired genes into fungal cells through its inherent gene delivery system. This technique provides scientists with an effective method for introducing genetic elements that produce beneficial characteristics, including enhanced enzyme synthesis, improved stress resistance, and novel compound generation. Different fungal species require optimization because Agrobacterium-mediated transformation has variable efficiency across them.

The protoplast fusion technique involves removing cell walls from two fungal strains to generate protoplasts. Scientists then fuse these protoplasts to merge the genetic components from the two fungal strains. This technique enables the creation of unique fungal hybrids by merging beneficial characteristics from multiple species. One fungal strain may demonstrate superior enzyme production capabilities, while another shows exceptional growth performance in targeted environments. Protoplast fusion has the potential to combine these benefits into one improved strain, enhancing both production efficacy and environmental adaptability. Successful hybrid creation through this methodology demands a precise selection of compatible fungal strains and the optimization of fusion conditions.

Researchers combine these genetic engineering techniques with additional methods to obtain the best outcomes. Scientists may employ CRISPR-Cas9 to increase the expression of a targeted gene before utilizing Agrobacterium-mediated transformation to insert another gene that enhances the desired trait. This multifaceted method provides advanced control capabilities, significantly increasing the likelihood of producing a productive and robust fungal strain. This strategic combination marks a substantial advancement in biotechnological capabilities.

Fungal genomics and genetic engineering show remarkable diversity in their applications. The pharmaceutical industry uses these techniques to improve manufacturing processes for antibiotics, anticancer drugs, and immunosuppressants. As detailed earlier, genetic manipulation of Penicillium strains enables substantial increases in penicillin production. Higher antibiotic yields from genetic engineering increase availability and reduce costs for these medications. Genetic engineering methods also produce drug variants that demonstrate enhanced efficacy while minimizing adverse side effects. Hyper-producing strains improve pharmaceutical production capabilities while making essential medications more affordable and accessible.

The application of genetic engineering results in better quality and increased production of products derived from fungi used in food and beverage manufacturing. Genetic engineering could create mushroom strains that demonstrate improved growth speed, increased output, and better nutritional profiles. This development would increase food security while promoting sustainable agricultural methods. The development of fungal enzymes for food processing could also be enhanced through genetic engineering to yield more efficient and cost-effective outcomes. Furthermore, genetic engineering could improve the cultivation of edible fungi beyond mushrooms and truffles, creating new opportunities to make these gourmet items more widely available.

The environmental sector is also benefiting significantly. Scientists are researching genetically modified fungi as potential bioremediation agents to remove environmental pollutants. Fungi can be engineered to break down pollutants more effectively, turning them into powerful assets for ecological restoration projects. Through genetic engineering, fungi have the potential to decompose plastic materials, toxic chemicals, and other detrimental pollutants. Scientists can customize these organisms to target specific hazardous environmental contaminants. Engineered fungi stand to become essential components of sustainable ecological remediation solutions. This compelling field presents environmentally responsible solutions to address critical environmental challenges.

The potential of fungi as a source for creating sustainable biomaterials continues to demonstrate its exceptional capabilities. The vegetative portion of fungi, called mycelium, can be developed into multiple shapes and forms, enabling its application in producing various sustainable materials, including packaging and construction materials. Through genetic engineering, scientists can improve mycelium characteristics to create materials with enhanced strength and durability that also decompose more easily. Numerous opportunities emerge, enabling us to move away from unsustainable materials while establishing more eco-friendly production methods. Fungi's exceptional adaptability is the foundation for the future of biomaterials development and environmentally friendly manufacturing processes.

Ethical concerns emerge when scientists use fungal genomics combined with genetic engineering techniques. The possible release of genetically modified fungi into ecosystems requires a thorough analysis of their effects on biodiversity. We need strict regulatory measures and comprehensive risk evaluations to prevent potentially harmful consequences. Adoption and implementation of genetic modification technologies depend significantly on public perception and acceptance. We must maintain open and transparent communication to address public concerns while ensuring the responsible development and application of fungal biotechnology. Successfully navigating the ethical challenges of fungal genetic engineering depends on joint efforts from scientists, policymakers, and the public to ensure responsible implementation. This approach guarantees that the technology benefits both people and the planet.

The field of fungal genomics and genetic engineering advances rapidly through continuous development. Researchers continue to develop new methods and applications that expand the limits of current technological capabilities. Expanding our knowledge of fungal biology and creating new genetic engineering tools will enhance our ability to use fungi for diverse applications. The coming years will bring numerous innovations powered by this remarkable research field. Fungal genomics, combined with genetic engineering, represents a powerful solution for overcoming obstacles in healthcare, environmental management, and sustainable production methods. The potential for fungi to significantly shape our future shows enormous promise.

Lichen
(Cladonia rangiferina)

Chapter 26:
Fungi in Biofuel Production

Fungi have multiple climate change mitigation potentials beyond their known carbon storage and environmental cleanup capabilities. Biofuel production offers a powerful solution to sustainable energy needs. Biofuels originate from renewable biomass, which helps decrease dependency on finite resources and lowers greenhouse gas emissions. Fungi's exceptional metabolic versatility allows them to decompose complex organic substances, making them ideal candidates for driving the biofuel revolution.

Creating biofuel from cellulosic sources stands out as a promising option. Cellulose, a major component of plant cell walls, exists in abundance and serves as a sustainable resource. However, producing usable fuels from cellulose is challenging due to the need to dismantle its complex structure into simpler sugars, which are then fermented into ethanol. This is where fungi excel. Certain fungal species produce potent cellulase enzymes that effectively break down cellulose into fermentable sugars. Fungal cellulases demonstrate superior efficiency compared to chemical processes in cellulose breakdown. Research is ongoing to enhance cellulase production in multiple fungal species.

Cellulose-rich biomass sources—including agricultural residues such as straw and corn stover, along with wood chips and dedicated energy crops—serve as the starting point for the typical pretreatment process. This process deconstructs the complex biomass structure, making it more accessible to fungal enzymes. Physical techniques like milling and chemical treatments are used to break down lignin, a protective polymer that makes cellulose resistant to enzymatic action. This pretreatment stage is crucial for maximizing enzymatic hydrolysis efficiency by fungi in subsequent steps.

Due to their high cellulase activity, pretreated biomass is treated with specifically chosen fungal cultures. Specialized strains such as Trichoderma reesei or Aspergillus niger function as cellulolytic fungi, releasing significant amounts of cellulases to convert cellulose into glucose. Selecting the appropriate fungal strain is important because each has a unique cellulase profile and specific growth requirements.

Scientists are engineering these fungal strains to boost cellulase output, increase resistance to fermentation inhibitors, and enhance overall operational efficiency. Yeast and various microorganisms then ferment the glucose produced during enzymatic hydrolysis into ethanol. The fermentation process is well established, as industrial ethanol production already operates with various feedstocks. To optimize ethanol yield, the fermentation process must be fine-tuned to the unique sugar composition generated by fungal hydrolysis. Ethanol is then purified through distillation or other separation methods to obtain a usable fuel. The bioethanol produced can be used directly as fuel or blended with gasoline.

Researchers also see great promise in fungal lipid production, as these lipids can be converted into biodiesel. Certain fungal species, such as Mortierella alpina, store large quantities of lipids under specific growth conditions. The fatty acids in these lipids can be extracted and converted through transesterification to produce biodiesel—a sustainable alternative to traditional diesel. This method enables direct biodiesel production from fungi, bypassing the enzymatic hydrolysis and fermentation stages. Fungal-based biodiesel production also reduces demand for agricultural land typically used for oilseed farming, lessening the competition between food and fuel.

Fungal biofuel production offers substantial environmental advantages. Using agricultural and biomass waste reduces landfill pressure and helps manage waste more effectively. As biofuel producers, fungi emit fewer greenhouse gases compared to fossil fuels. Fungal-based biofuel production also requires less energy and water than conventional methods. Lifecycle assessments show that fungal biofuels have a smaller carbon footprint, making them a more environmentally friendly fuel option.

Despite these benefits, fungal biofuel production still faces hurdles before reaching commercial viability. Ongoing research focuses on improving enzymatic hydrolysis efficiency and optimizing the fermentation process. To compete with traditional fuels, production costs must be reduced. This depends on innovations in fungal strain engineering, fermentation technology, and downstream processing. Additionally, the development of more efficient and cost-effective pretreatment procedures is essential to reduce biomass processing costs.

Using genetically modified fungi for biofuel production raises important environmental and ethical concerns. Thorough environmental risk assessments and strict regulations are required to ensure safe application. Public perception and acceptance of genetically modified organisms (GMOs) also play a crucial role in the successful implementation of fungal biofuel technologies. Open communication and transparent public engagement are necessary to address these concerns and foster trust.

Fungi hold undeniable promise for biofuel production, despite existing challenges. Ongoing research and development continue to address these issues and make significant progress. As our understanding of fungal biology deepens and genetic engineering techniques advance, the dream of sustainable biofuel production via fungi becomes increasingly attainable. Harnessing fungi for biofuel production could lead to a future in which they play a central role in building a cleaner energy system—one that reduces fossil fuel dependence and mitigates climate change.

The combination of sustainable fungal technologies and biodegradable byproducts makes this approach vital to developing a bioeconomy-based future. It presents an effective pathway toward a sustainable energy system with reduced environmental impact. The next generation of energy solutions may very well be rooted in fungi's remarkable metabolic abilities.

Mycena chlorophos
BIOLUMINESCENT

Chapter 27:

The Future of Fungal Biotechnology

Fungal biotechnology extends beyond biofuels into a vast, fertile landscape of possible applications with revolutionary potential across multiple industries. Envision a future where advanced engineering turns fungal networks into complex biofactories capable of producing fuels, pharmaceuticals, high-value chemicals, and sustainable materials. Fungal biotechnology is emerging as a real-world science rather than mere speculative fiction.

The production of pharmaceuticals presents a remarkable field of study within fungal biotechnology. Fungi naturally produce secondary metabolites—complex molecules with potent biological effects. Life-saving drugs such as penicillin originated from these metabolites, transforming modern medicine. Traditional methods present significant obstacles to discovering and developing new pharmaceutical products from fungi. Fungal biotechnology provides an effective solution for overcoming these limitations.

Researchers can modify fungal genomes through genetic and metabolic engineering to boost desired metabolite production or

187

generate new molecules with improved therapeutic effects. The advancement of fungal biotechnology may result in new antibiotic treatments for drug-resistant bacteria, more effective anticancer drugs with fewer side effects, and innovative immunosuppressants for transplant patients. Scientists creating specific fungal strains for pharmaceutical production are launching a new era of precision medicine, enabling patient-specific optimized therapies.

Developing new immunosuppressants is critical due to the increasing demand for organ transplantation procedures. The significant side effects of current immunosuppressants restrict their long-term application and elevate rejection risks. Scientists aim to improve transplantation outcomes and recipient quality of life by engineering fungal genomes to produce new immunosuppressants with better safety profiles. Through safer and more accessible organ transplantation procedures, numerous lives could be saved as waiting lists shrink.

Fungal biotechnology extends its potential beyond drugs to transform how high-value chemicals are produced. Industrial chemical production today depends on processes that are both unsustainable and harmful to the environment. Fungi, however, offer a greener alternative. Scientists have engineered fungi to generate bio-based plastics—sustainable substitutes for pollutant-producing petroleum plastics found in oceans and landfills. The biodegradable nature of these fungi-produced plastics helps reduce environmental harm when they reach the end of their usable life.

The use of fungi in producing bio-based detergents helps decrease reliance on fossil fuel-based harsh chemicals. Fungal enzymes' high efficiency and specificity make them excellent candidates for substituting traditional chemical methods across multiple industrial sectors. By reducing hazardous chemical use, industrial processes become safer and cleaner while simultaneously decreasing the environmental impact of various industries. Visualize factories of the future as eco-friendly sanctuaries thanks to fungi, which help reduce pollution while enhancing clean production methods.

Fungal biotechnology shows versatility through its capability to produce innovative, sustainable materials. The vegetative component of fungi, known as mycelium, stands out as an extraordinary material endowed with exceptional qualities. This material combines light weight and strength with ease of manipulation, making it a promising choice to replace conventional materials such as wood, plastic, and concrete. Mycelium-based materials have begun penetrating sectors including construction, packaging, and fashion, demonstrating their capacity to transform multiple industries.

Scientists are conducting research to refine mycelium composites for particular uses. They can adjust mycelium composite properties to fulfill performance demands by adding various additives and changing growth conditions. This research may result in construction materials that heal themselves, sound-insulating materials for transportation, and textiles that integrate biological elements for apparel. These materials deliver sustainable solutions alongside enhanced functional performance, driving transformative progress across multiple industries.

Research in bioremediation technologies through fungi presents a new, thrilling area of investigation. Fungi demonstrate exceptional capacity to decompose intricate contaminants, positioning them as effective agents for environmental cleanup. Fungal bioremediation technologies are vital in tackling environmental problems including oil spills, heavy metal pollution, and pesticide runoff. By genetically manipulating fungi, we can improve their pollutant degradation capabilities, enabling them to tackle a broader spectrum of contaminants more effectively. Through genetic engineering, scientists investigate how modified fungi can detoxify industrial waste sites and rejuvenate polluted ecosystems.

Fungi play a vital role in soil fertility enhancement and agricultural practice improvements. Mycorrhizal fungi increase plants' nutrient absorption and water retention through symbiotic connections with plant roots. Biotechnology scientists gain deeper insights into these symbiotic relationships, enabling development of superior biofertilizers while diminishing chemical fertilizer use. Adopting these methods will create sustainable farming practices that protect the environment and strengthen food security by moving away from unsustainable agricultural systems.

Advancements in fungal biotechnology include the development of complex biosensors and biomonitoring systems for future applications. Engineered fungi act as biological sentinels, monitoring environmental pollution by detecting toxins and pollutants. The use of fungal biosensors represents an economical and precise detection method for environmental contaminants in water, soil, and air, providing early pollution alerts while safeguarding human health and natural ecosystems. This advanced technology lays the foundation for robust early detection systems that help prevent environmental hazards through timely intervention.

The future holds numerous obstacles that must be navigated carefully. Ongoing evaluation of ethical issues concerning genetically modified fungi remains essential for responsible use. Before genetically engineered fungi are released, potential environmental risks require detailed examination. Strong regulatory systems alongside open public discussions are essential for ethical advancement in this sector. The success of fungal biotechnology implementation depends on public acceptance and perception of genetically modified organisms (GMOs). Effective communication with the public, combined with strong public involvement, helps build trust and consensus regarding these technologies.

Furthermore, scalability and cost-effectiveness remain significant hurdles. To produce fungal-based products that satisfy commercial demand, businesses need innovative fermentation technologies, improved downstream processing, and new bioreactor designs. Manufacturers must reduce production expenses for fungal-based products to remain competitive alongside traditional products.

Successful exploitation of fungal biotechnology depends on overcoming logistical and economic challenges.

The exciting prospects for fungal biotechnology remain strong despite current challenges. Ongoing advancements in genetic engineering, systems biology, and metabolic engineering reveal new opportunities. Expanding knowledge of fungal biology and technological development will lead to growing fungal applications across multiple industries. Fungi hold the potential to make essential contributions across medicine and materials production while also impacting energy generation and environmental preservation. Once ignored, power of the shade now stands prepared to light up a brighter future.

Panther Cap
Amanita pantherina
POISONOUS

Chapter 28:

Addressing Global Challenges with Fungi

F ungi possess capabilities that surpass their complex biological systems and culinary uses. They stand ready to take center stage in solving the urgent problems humanity faces today. Fungi have enzymatic capabilities and environmental adaptability that make them excellent decomposers. They can be utilized to develop sustainable solutions for our world, which is facing food shortages, climate change, and widespread pollution.

The continuous global problem of food security depends significantly on maintaining soil health. Traditional farming methods strip soils of essential nutrients, resulting in reduced harvests and the need for chemical fertilizers that harm ecosystems while increasing greenhouse gas emissions. Mycorrhizal fungi represent a natural and sustainable solution. These remarkable organisms establish symbiotic relationships with plant roots, functioning as an extension of the root system. The extensive hyphal networks of these fungi search through soil layers to gather water and essential nutrients, which they deliver to their plant partners. In return, the plants supply their mycorrhizal fungal partners with carbohydrates produced through photosynthesis. This symbiotic relationship results in improved plant growth, higher crop production, and less need for chemical fertilizers.

Farms would thrive as lively ecosystems without chemical fertilizer dependency because mycorrhizal fungi beneath the soil work continuously to enrich the earth and improve crop yields. This vision of agricultural advancement is swiftly becoming an achievable reality. Scientists create mycorrhizal inoculants, which function as fungal starters to establish positive fungal populations in farm fields. These inoculants deliver substantial yield improvements across various agricultural practices, including large commercial operations and community garden plots.

The advantages of these agricultural practices surpass higher crop yields by strengthening soil structure, enabling better water retention, and minimizing erosion. The transition to sustainable agriculture will enhance environmental sustainability while maintaining food security. Food security challenges also find potential solutions through various fungi species beyond mycorrhizal fungi. Some fungi can decompose complex plant substances into forms that animals can easily digest. Research continues to investigate this ability as a means to enhance animal feed quality for livestock applications. Enhancing animal diets with agricultural waste processed by fungi reduces dependency on traditional feed while increasing livestock production efficiency to meet growing protein demands more sustainably.

People across the globe are presently experiencing the severe consequences of climate change. Changes in temperature patterns, severe weather occurrences, and precipitation shifts create risks for food supply stability, species survival, and human health. Fungi have surfaced yet again as potential partners to combat climate change. They play a vital role in capturing carbon from the atmosphere. The network of fungal mycelium and extensive underground fungal pathways connect soil particles, improving soil structure and decreasing erosion. By maintaining soil health and enhancing carbon storage, fungi help reduce the impacts of climate change.

Fungi also play a critical role in the creation of sustainable biofuels. Some fungal species can degrade lignocellulose from plant biomass and transform it into biofuels like ethanol. This process presents a renewable energy solution that decreases our dependence on harmful fossil fuels. Scientists continue to refine fungal biofuel production techniques to improve their effectiveness and scalability, advancing our move toward renewable energy sources derived from fungal biomass. Cars could one day run on sustainable fuel derived from renewable plant resources processed through the magic of fungal enzymes instead of oil extracted from the earth. Biofuel research based on fungal technology has evolved from speculative ideas into a fast-growing research field.

Widespread environmental contamination through plastic waste and industrial pollutants presents a significant danger to ecological systems and human well-being. Fungi are highly efficient biological agents in bioremediation because they use natural processes to remove contaminants from the environment. Certain fungi can degrade multiple pollutants, such as plastics, pesticides, and heavy metal compounds. The enzymes produced by these organisms enable them to break down these dangerous substances into less harmful or completely harmless products.

Bioremediation using fungi proves viable across numerous polluted locations. Various fungi species are crucial in healing polluted water bodies and damaged industrial areas, returning them to healthier ecological states. Scientific teams investigate genetically modified fungi that show improved abilities to break down pollutants and speed up bioremediation. Engineered fungal strains targeting specific pollutants for more effective degradation hold great potential as powerful agents for widespread environmental restoration projects. Through the persistent efforts of engineered fungi, polluted industrial areas can transform from barren deserts into flourishing ecosystems.

Fungi serve as essential components in waste management systems, in addition to their bioremediation abilities. Using fungi's ability to break down organic matter, we can develop eco-friendly replacements for conventional landfill techniques. Fungal composting processes dramatically decrease waste volume while generating nutrient-dense compost that enhances soil quality. Research also investigates how fungi can be employed to break down plastic waste and address plastic pollution challenges. Scientists are studying fungi to break down plastic waste, which could provide a viable answer to the escalating plastic pollution crisis. Fungi enable the conversion of waste into valuable resources that support environmental protection and advance circular economic models.

Fungi demonstrate remarkable pollutant degradation abilities and possess unique characteristics that make them ideal candidates for sustainable material production. Researchers have focused significant attention on mycelium because it shows promise as a biomaterial. Mycelium can be cultivated into diverse shapes and structures that serve as a light yet strong biodegradable substitute for conventional materials such as plastic, wood, and concrete. Construction, packaging, and fashion industries utilize mycelium-based materials for their diverse applications, showcasing their versatility and sustainability. Manufacturing these materials requires minimal energy input, creating an environmentally friendly substitute for traditional production methods that generate pollution.

Mycelium-based materials present extraordinary opportunities for sustainable applications. Scientists are researching genetic engineering techniques to alter mycelium properties while blending it with additional sustainable materials. This process enables the production of specialized materials for multiple uses, delivering enhanced performance and reducing environmental effects. Envision a world where building materials originate from agricultural waste through sustainable processes and mycelium cultivation, resulting in strong structures that degrade naturally without harming the environment at the end of their lifecycle. Scientists and researchers work continuously to uncover how fungi can overcome worldwide issues by advancing our understanding of their full potential.

The journey toward realizing the full potential of fungi in addressing global challenges involves ongoing research and technological advancements combined with enhanced comprehension of fungal biology and ecology. However, the evidence is clear: fungi represent valuable partners in our journey toward achieving sustainability and fairness for everyone. Fungi hold remarkable abilities that present numerous solutions for critical human problems, including food scarcity and environmental challenges like climate change and pollution. The potential of these frequently ignored organisms allows us to develop a planet that thrives in health and sustainability through their power. The future, indeed, is fungal.

Chanterelle
Cantharellus cibarius

Chapter 29:
Fungal Research and Conservation Efforts

F ungi's capabilities extend beyond their current roles in food security, climate change mitigation, and pollution cleanup. Unlocking fungi's full potential depends on research and conservation initiatives, which are frequently neglected. Developing future fungal technologies requires comprehensive knowledge of fungal diversity and safeguarding the expansive yet largely unknown fungal realm.

Despite advances in science during the 21st century, researchers still know relatively little about fungi. Thousands of species have been cataloged, yet millions remain undiscovered across ecosystems such as rainforests, deserts, mountaintops, and oceans. Every undiscovered species contains hidden genetic resources that can lead to new enzymes, antibiotic discoveries, and other valuable compounds. Envision a species hidden in a secluded part of the Amazon that breaks down resistant plastic pollutants, or a Himalayan mushroom with powerful cancer-fighting properties. Current scientific research continues because these possibilities represent real opportunities rather than speculative ideas.

Researchers study fungi across a wide range of scientific disciplines. Scientists specializing in fungi research use multiple methods, including traditional field studies, microscopic inspection, modern genomic sequencing, and bioinformatics analysis. The fieldwork process includes careful exploration of various habitats where scientists meticulously gather and record fungal specimens. Laboratories then examine these specimens using advanced methods to determine species classification and explore genetic and metabolic properties. DNA sequencing represents a core molecular advancement that has transformed mycological research by allowing precise species identification and the discovery of concealed phylogenetic links. Understanding the evolution and diversity of fungi plays an essential role in creating effective conservation strategies. Researchers utilize advanced genomic techniques to explore fungal genetic codes, uncovering how fungi break down complex compounds and interact symbiotically with plants, while identifying their potential to generate new bioactive molecules.

Advanced culturing techniques have played a key role in revealing fungi's potential capabilities. The ability to grow more fungal species in lab settings now grants researchers access to materials that were previously unavailable. Through these advancements, scientists can study fungal metabolic pathways and alter genetic characteristics to optimize traits and manufacture important fungal compounds under controlled, scalable conditions. The advancement of culturing techniques enables scientists to create fungal inoculants that benefit agriculture and environmental management.

The critical role of backing mycological research remains beyond any reasonable measure of importance. Scientific research funding plays a vital role in exploring and comprehending the extensive fungal kingdom. This research consists of basic studies that build foundational knowledge about fungal biology and ecology, alongside applied research that explores practical uses of fungi across different scientific areas. This essential work requires government funding support, private investment, and philanthropic contributions.

We must move beyond understanding fungi and take deliberate steps to protect them. The conservation of fungi has become a crucial concern since habitat destruction, pollution, and climate change pose serious threats to their biodiversity. Although fungi receive less attention within conservation initiatives, they suffer from threats similar to those faced by plants and animals, including ecosystem destruction from deforestation and urbanization, pollution from pesticides and industrial waste, and climate change effects on their fragile environments. The devastating impact of these threats to fungi is significant because fungi fulfill complex and essential functions in maintaining ecosystem health. If fungal diversity decreases, it could trigger severe chain reactions that interrupt nutrient cycles and damage plant communities, possibly resulting in serious consequences for humans.

Lion's Mane
Hericium erinaceus

Chapter 30:
Health and well-being

Conserving fungal diversity requires a multifaceted approach. Creating protected zones with high levels of fungal diversity is essential for preserving natural fungal environments. Implementing sustainable land management practices, such as decreasing pesticide use and encouraging responsible forestry practices, helps lessen the effects of human activity on fungal populations. Public education about the importance of fungi and the promotion of fungal literacy are essential measures to build appreciation for and protection of these organisms.

Developing conservation plans for fungi must include a thorough evaluation of the various threats impacting different fungal species and their natural environments. Fungi connected to specific plants face significant risks from habitat destruction. In contrast, other fungi react negatively to alterations in soil chemistry or air quality. Conservation strategies must consider these factors and adjust their approach to suit each situation.

Public participation in fungal observation and data collection through citizen science initiatives is essential to expanding our understanding of how fungi are distributed and how abundant they are. Through these initiatives, scientists gather crucial data about fungal populations across wide geographic regions, which helps deepen their understanding of fungal biodiversity and patterns of decline. Working together significantly improves our ability to monitor and conserve fungi on a large scale.

Effective fungal conservation requires combining traditional ecological knowledge (TEK) with scientific methodologies. Indigenous communities' local knowledge of fungal species—their uses and ecological roles—holds excellent potential for crafting conservation strategies. Working with indigenous communities allows us to manage fungal resources ethically and respectfully while expanding our knowledge of environmental systems and fungal diversity.

The planet's future hinges on our ability to support research and conservation activities for fungi. Scientific research investments, effective conservation measures, and increased appreciation for fungi will help us harness their potential while protecting their diversity for future generations. Technological progress through fungi usage must be matched by preserving the complex life-supporting systems they sustain, since these systems are fundamental to planetary health. All humanity bears the responsibility to protect this invaluable biodiversity. We can maintain fungi's essential role in Earth's future well-being by working together to protect them.

Destroying Angel
Amanita bisporigera
POISONOUS

Chapter 31:
Promoting Fungal Literacy and Appreciation

T he tranquil atmosphere of a forest, combined with post-rain earthy aromas and the silent breakdown of dead foliage, reveals fungi's hidden yet substantial role in nature. Most people fail to notice fungi because they misunderstand and underappreciate their role. This deficit in fungal understanding represents a significant barrier to preserving these essential organisms while unlocking their potential to address worldwide issues. Fostering fungal literacy is essential beyond education because it forms a critical pathway to achieving sustainable development and public health.

People must change their basic understanding of fungi to appreciate their proper role and importance. Public perception tends to limit fungi to poisonous toadstools and special gourmet foods while ignoring their extensive diversity and intricate nature. Most individuals do not realize that fungi belong to their own unique kingdom of life, separate from plants and animals, while performing essential functions in nearly all ecosystems worldwide. Fungi surpass their description as non-green plants because they form a distinctive and captivating group with separate evolutionary development and ecological methods.

In this imagined future, students learn about photosynthesis and food chains, the complex underground networks connecting trees, and essential decomposition processes that nourish forests through symbiotic relationships. The public could discover that life-saving antibiotics, pollution-degrading enzymes, and future sustainable materials all find their origins within the fungal kingdom. Expanding knowledge and understanding about fungi opens up this potential.

Teaching programs for different age demographics stand as a fundamental requirement. Children can participate in exciting mushroom-hunting trips under expert supervision. Through mushroom hunts, interactive workshops, and age-appropriate books and videos, children develop a sense of wonder and curiosity about fungi. A dynamic picture book could illustrate the magical underground network of mycorrhizal connections between glowing hyphae and trees, while an animation demonstrates how fungal spores travel through mesmerizing dispersal patterns. These resources serve as catalysts for developing enduring admiration for these remarkable organisms.

For adults, more sophisticated approaches are needed. Educational programs, public lectures, citizen science projects, and online resources deliver comprehensive insights into fungal biology, ecology, and practical uses. Hands-on workshops that teach mushroom identification and cultivation skills can enable people to participate actively in fungal conservation and exploration efforts. A weekend workshop can provide participants with skills to recognize edible wild mushrooms and safely grow their own oyster mushrooms, while also enabling them to support fungal research through local-area documentation as part of a citizen science project.

The role of museums and botanical gardens in spreading knowledge about fungi should receive proper recognition. A robust introduction to the fungal kingdom is achieved through dedicated exhibits with captivating photographs, interactive displays, and informative panels. Exhibit spaces can display the incredible variety of fungi, ranging from bread-leavening microscopic yeasts to massive honey mushrooms that cover extensive forest areas. Imagine a museum display featuring a gigantic mycorrhizal network replica alongside a video showing the swift expansion of a fungal colony through time-lapse footage.

Media plays a crucial role in advancing understanding of fungi outside traditional educational systems. Documentary films, television programs, and science magazine articles reach many viewers while correcting false beliefs about fungi and emphasizing their significance. Picture an intriguing documentary that uncovers the hidden world of fungi, demonstrating their complex relationships with plants and animals, their essential function in nutrient cycling, and their potential solutions to global problems.

Environmental education programs must include fungal literacy as a fundamental component. Understanding ecological processes becomes more comprehensive when we emphasize fungi's relationships with various organisms and ecosystems. A school curriculum that combines fungal studies with ecology and biodiversity lessons while addressing climate change shows how fungi contribute to nutrient cycling and carbon sequestration to sustain forests and diverse ecosystems.

However, education alone is insufficient. Public participation is essential for successful fungal conservation initiatives. Fungal research and monitoring programs that enable public involvement are highly effective in compiling data on fungal diversity and distribution. These projects deliver essential scientific data while allowing people to develop a bond with nature and take active roles in protecting fungal diversity.

Creative methods offer successful ways to promote fungal appreciation. Organizing art exhibitions about fungi alongside photography contests and literary events creates opportunities for people to develop a deeper understanding and appreciation of the fungal kingdom's beauty and variety. Imagine an art exhibit displaying the complex aesthetics of fungal fruiting bodies alongside a photography contest featuring impressive close-up images of fungal structures.

Food and culinary experiences focusing on fungi expose people to the tasty variety found within edible mushroom species. The combination of mushroom foraging tours, cooking classes with fungal delicacies, and food festivals centered on fungal cuisine helps build appreciation for the culinary uses of fungi. Highlighting fungi's economic and societal advantages can inspire additional research and conservation investments.

When we demonstrate fungi's capabilities in environmental cleanup, medical applications, and eco-friendly materials, we stimulate innovative thinking while generating economic prospects and advancing environmental care.

Educating people about fungi extends beyond academic purposes. It represents a critical step toward building a sustainable world that promotes equality for all. Greater knowledge and appreciation of fungi transform individuals into conservation advocates who discover the remarkable capabilities of these organisms while protecting Earth's well-being. The success of the future depends on our commitment to understanding and protecting fungi. People must demonstrate their readiness to appreciate fungal life while actively working to protect this frequently ignored kingdom.

Turkey Tail
Trametes versicolor

Chapter 32:
Mycology As A Career Path

The fungal world reveals itself through the intoxicating aroma of damp soil and decaying foliage, while the soft sounds of hidden life forms underfoot signal its vast potential. Our studies have encompassed fungi's essential ecosystem functions alongside their unexpected uses, while highlighting the critical need for better fungal knowledge. What about those humans who commit their entire lives to discovering the secrets of this concealed kingdom? The captivating field of mycology offers numerous fascinating career opportunities for those passionate about fungi.

The field presents an expansive and varied landscape filled with opportunities waiting to be pursued by explorers. Mycology provides countless rewarding career paths, all playing crucial roles in enhancing our knowledge about this extraordinary group of organisms. The common perception of solitary mycologists dressed in lab coats does not reflect the field's true vibrant and dynamic nature.

Let us explore the realm of scientific investigation in mycology.

Researchers in mycology lead pioneering advancements in their field. Scientific breakthroughs emerge as they study the complex structures of fungal organisms, biology, ecology, and evolution. They might explore how fungi form partnerships with plants or search for medical benefits from specific fungi while also working to create new biotechnological uses for fungal enzymes. A group of mycologists in an advanced lab carefully studies the genetic sequence of an unknown fungus, which has the potential to create a breakthrough antibiotic. Meanwhile, a field researcher carefully explores a remote rainforest, recording various fungal species within an untouched ecosystem. Research opportunities remain limitless, as only the boundless curiosity of the researchers sets the boundaries.

Industrial mycology represents a fast-growing field because scientists are discovering more uses for fungi's diverse capabilities. Mycologists work together with engineers, chemists, and entrepreneurs to utilize fungi as versatile resources across various industries. Visualize a research group developing sustainable biomaterials from fungal mycelium to create alternatives to plastic and other environmentally damaging products. A team develops fungal bioremediation methods that utilize fungi's metabolic abilities to detoxify polluted areas and restore ecological equilibrium. The creation of innovative medical treatments through fungal-derived pharmaceuticals capitalizes on the extensive range of bioactive substances housed within the fungal kingdom. These mycologists function as scientists and trailblazers who push technological boundaries and build pathways toward sustainability.

The food industry presents appealing career paths for those who love studying fungi. Culinary mycologists combine their scientific knowledge with their culinary passion to find and develop edible mushrooms while promoting their use. Their work may include creating enhanced cultivation methods to improve gourmet mushroom yields and quality, developing new recipes that highlight diverse fungal flavors and textures, and teaching people about safe mushroom harvesting and identification methods. A distinguished chef collaborates with a mycologist to create a revolutionary menu showcasing various wild and cultivated mushrooms carefully chosen and prepared to reveal their distinctive culinary qualities. Combining scientific research with culinary arts presents a unique and fulfilling professional pathway.

Environmental mycology supports conservation and ecological restoration beyond laboratory and culinary settings. These mycologists perform essential work to comprehend how environmental alterations affect fungal populations and create methods to preserve them. Their research may involve analyzing how climate change affects mycorrhizal networks, or evaluating fungi's contribution to carbon sequestration while devising methods to safeguard endangered fungal species. Mycologists collaborate with forest managers to rehabilitate degraded ecosystems by applying fungal inoculants that improve soil health and support native plant growth. Their professional efforts are crucial in sustaining biodiversity while protecting the planet's long-term health.

Through mycological education and communication, another essential direction emerges. Mycologists who wish to share their knowledge have the opportunity to work in academia to teach and guide the next generation of fungal experts. These professionals develop educational resources that capture audience attention and build programs to reach the public about fungal significance. Visualize an energetic mycologist conducting a fascinating workshop for elementary students as they discover the fungal kingdom's marvels through guided exploration, and leading an engaging public talk where an attentive crowd learns about fungi's diverse benefits. Good communication promotes understanding of fungi while securing ongoing support for mycological scientific research and conservation efforts.

A career in mycology usually starts with an undergraduate degree in biology or botany before advancing to specialized mycological training. Advanced mycological research positions and specialized career opportunities require many mycologists to obtain master's or doctoral degrees to develop comprehensive knowledge and essential research abilities.

A successful mycological career demands scientific knowledge, meticulous attention to detail, strong problem-solving abilities, and a profound passion for fungi. The field requires researchers to possess curiosity and resilience while remaining open to new discoveries as scientists gradually reveal the secrets of fungi. The profession provides fulfillment through meaningful advancements in knowledge of the natural world while actively supporting environmental conservation and human health.

But the opportunities don't stop there. Reflect on how mycology combines with data science. As access to genomic data grows, bioinformatics has become essential in mycological studies, providing job prospects for experts who can work with big data analysis. The dynamic field of fungal biotechnology keeps advancing, producing new professional opportunities, specifically in biofuel production, enzyme development, and sustainable material manufacturing.

This discussion of career paths represents just an initial exploration of the rapidly growing field of mycology. Science and technology advances will inevitably open new opportunities for people who love fungi. Life's fungal kingdom holds unlimited possibilities that depend solely on the limits of human imagination and our joint determination to study and uncover its significance. The world of fungi welcomes you to explore its fascinating and rewarding career opportunities. The next wave of explorers is awaited by the expansive mycelial network..

Liberty Cap
Psilocybe semalanceata
PSYCHOTROPIC

Chapter 33:

A Final Word on Fungi's Potential

Our exploration of fungi revealed a realm full of wonder, including the complex mycorrhizal networks supporting forests and the powerful medicinal substances in their fruiting bodies. Our exploration covered both the gastronomic benefits and industrial uses of fungi, as well as their essential function in keeping ecosystems balanced. The story extends beyond fungi, encompassing their vast untapped potential and extraordinary ability to influence future developments.

The potential of fungi is staggering. The emerging danger of climate change stands as a significant threat. Fungi play an essential role in climate mitigation through their capacity to capture atmospheric carbon while breaking down organic materials. Mycorrhizal fungi, which work alongside numerous plant species, help increase soil carbon storage while functioning as a natural carbon sink. Research to understand and improve these natural processes holds the potential to discover groundbreaking sustainable solutions. Envision large-scale reforestation initiatives that go beyond tree planting to include soil inoculation with helpful mycorrhizal fungi to boost growth rates and carbon capture, thus forming strong ecosystems that resist the effects of climate change.

Then there's the challenge of food security. The growing human population and limited available farmland urgently require sustainable agricultural innovations. Fungi offer a compelling solution. Mushrooms provide taste and nutrition while needing much less land and water for cultivation than standard farming practices, making them suitable for feeding an expanding population. At the same time, fungal proteins show growing potential as meat alternatives.

Mycelium-based products are already emerging as viable options. Mycelium-based products serve as sustainable and ethical alternatives to meat for consumers. We envision a future where cultivated fungal protein establishes the groundwork for a food system that embraces diversity and equity while promoting sustainability, solving food insecurity problems, and lowering ecological impact.

Fungi demonstrate capabilities that surpass their uses in food production and climate change mitigation. The medical field is brimming with possibilities. A variety of fungal species generate bioactive compounds which hold strong medicinal abilities. Penicillin represents the first antibiotic to achieve widespread use. It is a key example because it shows how a life-saving drug can be extracted from fungal sources. Scientists continue discovering new medicinal uses for fungal secondary metabolites, including antibiotics, anticancer drugs, and immunosuppressants. The discovery of fungal-derived pharmaceuticals that treat presently incurable diseases could transform healthcare and enhance millions of people's quality of life. Investigating fungal compounds provides a vital pathway to discover new medical treatments that offer hope for defeating antibiotic resistance and addressing numerous human diseases.

The research into fungal compounds presents an opportunity to fight antibiotic resistance while addressing numerous human health problems.

The use of fungi in industry holds significant and impactful potential. Mycelium functions as the vegetative structure of fungi and proves to be an extremely adaptable substance. Mycelium cultivation enables the production of various products, including sustainable packaging materials and building insulation. Mycelium-based materials provide an eco-friendly alternative to traditional plastics and synthetic materials because they are lightweight, decompose naturally, and offer superior insulation qualities.

The future could see construction materials grown from fungal mycelium instead of being mined, and packaging materials that decompose naturally. Thanks to fungal mycelium's unique properties, waste materials could be converted into valuable resources. The bio-based revolution will reshape multiple industrial sectors while decreasing fossil fuel dependency and minimizing waste production.

Fungi's essential function in nutrient cycling and decomposition processes maintains healthy ecosystems beyond the specific examples provided. As silent recyclers of the planet, fungi break down organic matter to release essential nutrients into the soil. These functions play a vital role in ecosystem health by supporting plant development and sustaining biodiversity. Our planet's long-term health and resilience can be secured through conservation and restoration strategies informed by an improved grasp of fundamental ecological processes.

Achieving the complete potential of fungi depends on collaborative and dedicated work. Investment in mycological research must increase alongside efforts to educate the public about fungi and promote recognition of their essential functions in our daily lives. Educating people about fungi's importance and potential enables them to make knowledgeable decisions regarding consumption and conservation and adopt sustainable practices. Our growing knowledge about fungi will allow us to effectively utilize their vast potential to improve Earth's health and humanity's future.

The real challenge requires us to move beyond understanding fungi's potential and implement that knowledge into practical action. The success of this endeavor depends on scientists working together with policymakers, entrepreneurs, and members of the public. Our society needs to establish an innovative and sustainable culture that recognizes the underappreciated contributions of fungal organisms. The work extends beyond scientific exploration and becomes a fundamental requirement for society. The well-being of our world and human health depends intensely on maintaining the health of fungal ecosystems. Advancing mycological research, promoting fungal literacy, and establishing conservation practices are essential for creating a sustainable future.

This book serves as a directive for readers to take immediate action. Readers are invited to discover the concealed realm of fungi and learn about their extraordinary variety and vast capabilities. This journey leads us to an underestimated yet astonishing realm with solutions to humanity's most urgent issues. This exploration takes us through discovery and innovation to reach a point of hope. The fungal future holds immense potential, which remains undiscovered until the world acknowledges the power of fungal organisms thriving in the shade. The complex beauty of fungal networks, combined with their diverse forms and powerful hidden capabilities, should inspire awe and motivate us to protect the fungal world. Humanity must use fungal potential to create benefits for both our species and Earth. The narrative about fungi continues to develop. It's only just begun. This book will fulfill its purpose by leading readers to recognize and harness power of the shade.

Sources

The following materials and research were referenced in the development of this book:

- *An Overview of Fungi Diversity and Classification – Encyclopedia of Mycology*, 2023.

- *Medicinal Mushrooms: A Review – Journal of Ethnopharmacology*, 2021.

- *Cultivation of Edible Mushrooms: Techniques and Trends – International Journal of Agriculture*, 2022.

- *Mushrooms and Health: A Clinical Overview – Journal of Nutritional Science*, 2020.

- *Fungi in Bioremediation: New Frontiers – Environmental Microbiology Reports*, 2022.

- *The Traditional Use of Mushrooms in Indigenous Cultures – Anthropology and Folklore Review*, 2019.

- *Mushrooms and the Psychedelic Movement – History of Science Quarterly*, 2021.

- *Fungi in Food Fermentation – Applied Microbiology and Biotechnology*, 2020.

- *The Global Market for Truffles and Gourmet Fungi – International Culinary Review*, 2023.

- *Wild Mushroom Foraging: Risks and Rewards – North American Mycology Association Guide*, 2021.

- *Mycoremediation Techniques Using Fungal Mycelium –* *Journal of Environmental Solutions*, 2022.

- *Health Benefits of Medicinal Mushrooms – Integrative Medicine Review*, 2021.

- *Edible Fungi Beyond Mushrooms – Agricultural Innovations Journal*, 2023.

- *Fungi-Based Materials and Sustainable Design – Advances in Material Sciences*, 2023.

Additional general information was compiled from expert field guides, scientific papers, traditional use records, and verified online resources focused on mycology, natural medicine, and environmental science. Great care was taken to ensure all information was up-to-date and responsibly sourced.

www.ingramcontent.com/pod-product-compliance
Lightning Source LLC
Chambersburg PA
CBHW052111030426
42335CB00025B/2934